Science and Religion

D0265947

Yves Gingras

Science and Religion

An Impossible Dialogue

Translated by Peter Keating

polity

First published in French as *L'Impossible dialogue. Sciences et religions* © Éditions du Boréal, 2016

This English edition © Polity Press, 2017

Polity Press
65 Bridge Street
Cambridge CB2 1UR, UK

Polity Press
350 Main Street
Malden, MA 02148, USA

All rights reserved. Except for the quotation of short passages for the purpose of criticism and review, no part of this publication may be reproduced, stored in a retrieval system or transmitted, in any form or by any means, electronic, mechanical, photocopying, recording or otherwise, without the prior permission of the publisher.

ISBN-13: 978-1-5095-1892-0
ISBN-13: 978-1-5095-1893-7 (pb)

A catalogue record for this book is available from the British Library.

Library of Congress Cataloging-in-Publication Data

Names: Gingras, Yves, 1954- author.
Title: Science and religion : an impossible dialogue / Yves Gingras.
Other titles: Impossible dialogue. English
Description: Malden, MA : Polity, 2017. | Includes bibliographical references and index.
Identifiers: LCCN 2017010097 (print) | LCCN 2017011427 (ebook) | ISBN 9781509518920 (hardback) | ISBN 9781509518937 (pbk.) | ISBN 9781509518951 (Mobi) | ISBN 9781509518968 (Epub)
Subjects: LCSH: Religion and science--History.
Classification: LCC BL240.3 G56413 2017 (print) | LCC BL240.3 (ebook) | DDC 201/.6509--dc23
LC record available at https://lccn.loc.gov/2017010097

Typeset in 10.5 on 12 pt Sabon by Servis Filmsetting Ltd, Stockport, Cheshire
Printed and bound in the UK by CPI Group (UK) Ltd, Croydon

The publisher has used its best endeavours to ensure that the URLs for external websites referred to in this book are correct and active at the time of going to press. However, the publisher has no responsibility for the websites and can make no guarantee that a site will remain live or that the content is or will remain appropriate.

Every effort has been made to trace all copyright holders, but if any have been inadvertently overlooked the publisher will be pleased to include any necessary credits in any subsequent reprint or edition.

For further information on Polity, visit our website: politybooks.com

CONTENTS

TABLES AND FIGURES

Tables

Figures

If there is one truth that history has incontrovertibly settled, it is that religion extends over an ever diminishing area of social life. Originally, it extended to everything; everything social was religious – the two words were synonymous. Then gradually political, economic and scientific functions broke free from the religious function, becoming separate entities and taking on more and more a markedly temporal character.

Emile Durkheim[1]

INTRODUCTION

While religions divide men,
Reason brings them closer.
Ernest Renan[1]

This book attempts to explain how the question of the relations between science and religion and calls for a 'dialogue' between these two areas of activity, so distinct in their objects and methods, came to occupy a significant place in public discussion in the course of the 1980s. For it was not always so. Until quite recently, the scientific consensus was, as a botanist and Brother of Christian Schools expressed it clearly in the mid-1920s, that 'science and religion follow parallel paths, towards their own goals', and that there was no need to search for a necessary 'harmony' between scientific discoveries and religious beliefs.[2] As a student in physics during the 1970s, I recall that neither students nor professors spent much time discussing the supposed relations between science and religion nor was there any spirited public debate or a plethora of books on the topic. Even during the 1980s, when studying the history and sociology of science, such discussions were still rare and largely limited to the counter-culture and followers of 'New Age' syncretic 'philosophies'. Thus the question: how did this renewed interest for a 'dialogue between science and religion' come about?

As will be seen in chapter 5, the return of these issues to prominence in the intellectual field has, as one of its sources, Pope John Paul II's decision in November 1979 to review the Galileo trial, a potent symbol in both popular and academic imagination of the opposition between scientific thought and religious beliefs. The mere mention of the condemnation of Galileo by the Holy Office (then known

1

as the Supreme Sacred Congregation of the Roman and Universal Inquisition) in June 1633 suffices to remind us that the relations between science and religion have an extensive history, and that, on many occasions, they have been quite fraught. While the 1979 decision made at the highest level of the Catholic hierarchy may have triggered interest in the issue, it does not entirely explain the proliferation of publications since the 1980s with titles that juxtapose the words 'science', 'religion' and 'God'. The 1960s and 1970s witnessed the rise of a current of syncretic thought associated with 'counter-culture' and 'New Age' seeking to combine philosophical and religious traditions with the 'mysteries' of quantum physics, a theory seen as a challenge to both logic and common sense. This in part created a platform for the many popular books that claimed that the 'the most advanced' science had now confirmed the intuitions of the 'oldest' spiritual traditions.[3] Since the publication of the physicist Fritjof Capra's seminal work, *The Tao of Physics* (1975), publishers have sensed a commercial opportunity, and the market has burgeoned with books relating God and science, bearing similarly eye-catching titles. That these associations of science and religion, for the most part superficial, are the result of sincere beliefs or mercantile cynicism is of little importance. What must be understood here, as will be seen in chapter 6, is how some scientific discoveries are used to justify religious and theological positions that have little to do with science. Instead, science's prestige is used to suggest to religious readers that modern science is in fact compatible with their preferred beliefs. Moreover, faced with the rise of fundamentalist religious sects – often highly critical of research that questions their beliefs – a number of scientists and their organizations, seeking appeasement, have come to support these dubious associations that suggest that believers need no longer be wary of modern science. Far from leading the innocent down the road to atheism, as is often thought, science leads instead, they surmise, to a belief in a nature created by a superior being.

Another important factor in the exponential growth of studies devoted to the study of the relationship between science and religion over the past twenty years has been the work of John Templeton (1912–2008) and his Foundation. As we will see in chapters 5 and 6, this Foundation, endowed with more than a billion dollars, distributes millions yearly to researchers who seek links between science, religion and spirituality. Since the mid-1990s, the Templeton Prize has frequently been bestowed upon astrophysicists who offer – directly or indirectly – religious or spiritual interpretations of modern physics and cosmology. The Foundation has also played a major role in foisting the theme of a 'dialogue' between science and religion onto the history of science. It will also be seen in chapter 6 that these so-called 'dialogues' are little more than a modern version of natural theology, employing arguments that have barely changed since the end of the seventeenth century.

But before analysing the rise of the discourse on the relationship between science and religion, we will examine the long history of their conflicting relations. For despite the recent trend among many historians of science to say that the conflict between science and religion has been largely exaggerated, it remains the case that many scientific theories have historically been perceived as incompatible with certain religious beliefs that are based on the literal reading of sacred texts. While it is true that these clashes of world views are in some respects contingent and become open conflicts only when organized social groups or institutions confront the offending science, it is also true that the clashes are often predictable and even inevitable when a given science takes on issues and problems that resonate with those discussed in 'sacred' religious writings. In sum, if mathematics or taxonomy pose few problems for organized religion, the same cannot be said for cosmology, geology, evolutionary biology and the social and human sciences, especially those that deal with the history of religion and the origins of humanity. As noted by the sociologist Max Weber at the beginning of the twentieth century, 'the tension between religion and intellectual knowledge definitely comes to the fore wherever rational, empirical knowledge has consistently worked to the disenchantment of the world and its transformation into a causal mechanism. For then science encounters claims of the ethical postulate that the world is a God-ordained and hence somehow meaningfully and ethically oriented cosmos'. He added, moreover, that 'the extent of consciousness or of consistency in the experience of this contrast, however, varies widely'.[4]

The undeniable historical conflicts between different sciences and various religions participated in a struggle for power between institutions and groups with divergent and often opposed interests. At the dawn of the development of modern science in the seventeenth century, the power and prestige of scientific institutions paled in comparison to that of the Christian Church that dominated the intellectual world. Having become a symbol of the history of the relationship between science and religion, the condemnation of Galileo in 1633 merits special attention in the first two chapters. Chapter 1 shows how the balance of power between scholars and theologians explains Galileo's condemnation. Though this history has often been told, we focus here on less discussed aspects related to the defence of the *autonomy* of astronomers' and natural philosophers' discourse vis-à-vis that of theologians. Chapter 2 then recalls the numerous attempts by philosophers and scientists in the three centuries that followed to overturn his condemnation and rehabilitate him, as well as the books of Copernicus and Kepler, which had also been condemned. This saga was finally put to rest by Pope John Paul II in 1992 on the 350th anniversary of the death of the Italian scientist. We will see in chapter 3 that although cosmology was long the source of many a bone of contention between science and Christian theology, it was replaced at the beginning of the nineteenth century by natural history and geology. Subsequently institutionalized, these sciences applied a naturalist methodology to the whole of nature with the result that invocations of the divine (or of miracles) no longer had a place in science by the middle of the nineteenth century. This long process of autonomization thus accompanied the relegation of God to the periphery of scientific discourse, even for scientists who, as individuals, remained fervent Christians.

Against the current trend – dominant in history of science since the end of the 1980s – that tends to deny or minimize the existence of significant conflicts between science and religions, chapter 4 reviews the many cases of censorship of the scientific literature by the Roman Church and its organizations (the Congregation of the Index and the Inquisition) from the beginning of the seventeenth to the middle of the twentieth century. It will also be seen that although the many Protestant sects were less organized than the Catholic Church, they were also able to prohibit publications and to dismiss or silence scientists whose scientific views contradicted their religious credos.

Although the Catholic Church, as an institution and an organization, has finally lost much of its temporal power, less centralized religious organizations (such as those of Protestants and Muslims) now act as pressure groups to block the teaching of theories they deem offensive (such as the theory of evolution), or research they deem immoral (such as stem cell research). Seeking to ban subjects offensive to their 'deep' convictions, they may take control of local school boards or adopt laws to limit the teaching of scientific theories such as evolution, as has happened in the most conservative states of the United States (since the 1920s) and in some Muslim countries (since the 1980s). They may also seek to impose the teaching of religious conceptions under the guise of science, as the long American debate around the notion of 'intelligent design' has shown.[5]

4

These days, however, the authority of science is pretty much undisputed and, as discussed in chapter 6, it is this very power that often obliges promoters of religious beliefs to seek legitimacy through science. Indeed, just as some scientific theories may be refused in the name of religion, others can be used to lend credibility to a given religious enterprise.

A brief remark on the vocabulary used in this book is needed to avoid the frequent and occasionally intentional confusion surrounding the meaning of the words *religion* and *science*. Although the sciences have multiple objects and methods, we will generally employ the singular – science – partly to avoid complicated and often pedantic formulations, but more especially to emphasize a point common to all contemporary science, be it the natural or the social sciences: *these activities are attempts to provide reasons for observable phenomena by means of concepts and theories that do not call upon supernatural causes*. This is often called 'scientific naturalism' or 'methodological naturalism', and it is a *postulate* of the scientific world view. The separation of science from religion has, of course, been gradual, as will be seen in particular in chapter 3 of this book. But while some historians suggest that we avoid speaking of conflict between 'science and religion', under the pretext that the content of these terms changes over time, this is but a pious wish. They themselves do not follow their own advice, since all the books on the subject, including theirs, use the words *science* and *religion*.[6] It should be obvious that under the stability of a word, there is often a variable content over the centuries. But that does not entail that there was no distinction between the two terms at any given time.

As regards 'religion', it is well known that there is no perfect definition that captures its content.[7] For the purposes of this book, which is to analyse the historical relations between science and religion as *institutions* in the western world since the seventeenth century, it suffices to distinguish private beliefs and spirituality on the one hand – which do not concern us here – from religions as public institutions on the other. As organizations that are more or less centralized, religions function through spokespersons (theologians, pastors, imams, etc.), who are the guardians, defenders and promoters of their respective dogmas and practices. Varying by period and region, and according to their modes of organization – centralized or not – these spokespersons have varying degrees of power (direct and indirect) over society, and hold greater or lesser control over individuals. The sociologist of religion Danièle Hervieu-Léger thus defines religion as a 'practical and symbolic ideological device, by which the consciousness (individual and collective) of belonging to a line of belief is incorporated, maintained, developed and controlled'.[8] It is of course the control over science and its autonomy that concerns us here. And before becoming an object of study for the social sciences, 'religion was [its] opponent', and it is the 'fight for the secular autonomy of knowledge', to quote again Hervieu-Léger, that historically constituted the basis for the formation of relatively autonomous scientific communities.[9]

Modern science, sometimes referred to as 'western science', as it emerged in Italy and Northern Europe (England, France, Netherlands and the German-speaking world), developed primarily in the space defined by Christianity. It is therefore primarily the Christian Church that will be at issue in this book. The Muslim world has not played a central role in the development of modern science since the seventeenth century.[10] The conflicts generated by the emergence of modern science thus largely implicate Christianity. In its institutional manifestations, this encompasses the various Protestant churches from the Reform and of course the Catholic Church and its armed wings: the Inquisition and the Holy Office.[11] Occasionally, debates such as those concerning Darwinism in the second half of the nineteenth century were exported into the Muslim world via educational institutions. For example, the colonial Syrian Protestant College (since 1920: the American University of Beirut), was founded in Beirut by American evangelical Christians in the middle of the 1860s. A clear expression of the invariant character of the discourse on the relationship between science and religion can be found in the fact that the young Muslim intellectuals trained in the Syrian Protestant College would go on to reformulate primarily 'western' arguments for both the harmony and the conflict between science and the Koran.[12] But it was not until the 1980s, in the context of the political and national affirmation of the oil-producing countries, that we observe a surge in Islamic criticisms of the theory of evolution.[13]

As the Korean theologian Seung Chul Kim rightly pointed out, the many works falling under the heading 'science and religion' concern primarily the relationship between science and Christianity.[14] More generally, it is doubtful that it even makes much sense to speak of 'religion' in the case of spiritual practices not based on a revealed text and a single God.[15] Pantheistic religions that identify their divinity with nature have had few problems with science *per se*. It is therefore hardly a coincidence that the cases of conflict between science and religion are found essentially in the Christian, Muslim and Jewish traditions. All three have constructed cosmologies based on texts considered sacred. Containing the words of God, they have given rise to literal interpretations that conflict with modern scientific discoveries.

* * *

By using an institutional approach, this book departs from the view that is currently dominant amongst historians of science who have studied the question of the relationship between science and religion since the mid-1980s. But this literature presents many methodological problems. Too many debates confuse scientists' personal religious convictions and the stance of the Church as an institution. There is a curious tendency to constantly downplay clear conflicts between scientific theories and particular beliefs promoted by religious institutions, to confuse individual and institutional levels of analysis, or even to remain silent on historical sources showing clear conflict between theology and science (we will provide examples of these processes in chapters 4 and 5). For instance, to show that the personal beliefs or religious motivations of a given researcher positively influenced his or her research may be interesting from a biographical point of view, but it fails to enlighten us about the way in which religious institutions have responded to a given scientific discovery or theory.[16] In this vein, the historian of science John Hedley Brooke, one of the most prolific contributors to the development of the 'industry' of the history of science–religion relations since the end of the 1980s, tells us that: 'serious scholarship in the history of science has revealed so extraordinary rich and complex a relationship between science and religion in the past that general theses are difficult to sustain'.[17] In addition to the fact that the rhetorical use of the qualifier 'serious' excludes from the outset as 'non-serious' any work proposing to give a more general picture, venturing beyond the tiny details that characterize the specific cases and the historical periods studied, the emphasis on 'complexity' renders institutions invisible and creates the illusion of an absolute contingency of events. It also allows one to avoid any statement that tends towards generality and even suggests that if you look closely enough, you can no longer truly distinguish science from religion, and vice versa.[18] However, issues invisible at the individual level come clearly into view at the scale of institutions. The insistence on the 'complexity' and 'richness' of the relations between science and religion – which some historians even call the 'complexity thesis' as if it were not obvious that history is always 'complex' and even complicated! – conjures up the image of a physicist who attempts

7

to understand the behaviour of a gas in a container by focusing on the trajectory of a single molecule. At this scale, it is clear that the molecule moves randomly in reaction to unexpected collisions with other molecules. But this microscopic chaos gives way, at a larger scale, to the relatively simple Boyle's law that relates the pressure to the volume and temperature of the gas.

Consequently, we must first identify the scale of analysis (individuals or institutions). We must also, as we will see in chapter 3, distinguish what philosophers of science call the context of discovery (or better, of the pursuit of research), from the context of justification. The first depends on the personal beliefs and convictions of each researcher, and may thus include a religious or spiritual dimension. The second refers to the way research results are discussed and dissected within the scientific community, and thus has an institutional dimension. For it is indeed institutions that set out the rules of the game, and that establish the legitimacy of the arguments acceptable to the scientific community at any given time. In the collective activity of science, knowledge is sanctioned by a community of researchers who define what is considered accepted, refuted or still subject to debate at any given moment of history. Individual religious beliefs do not constitute an *institutional* criterion of validity, even though they can obviously offer powerful motivation for some scholars. Both levels of activity can of course be studied by historians, but here we will focus on the second (institutions), which have long been neglected in favour of the first (individuals). Consider, for instance, the many publications on Einstein's supposed 'religion', even though this icon of physics was more pantheistic than religious and was quite averse to the simplistic belief in a personal god that can respond to specific requests. Indeed, in the same paper in which he wrote the oft-quoted remark that 'science without religion is lame, religion without science is blind', Einstein affirmed that: 'the main source of present-day conflicts between the spheres of religion and of science lies in the concept of a personal God' and that 'teachers of religion must have the stature to give up the doctrine of a personal God'.[19]

The rhetoric deployed in most of the recent publications on science and religion always minimizes well-known and high-profile conflicts and sometimes goes as far as unearthing the positive aspects of censorship. The most flagrant example is the condemnation of Aristotle's philosophy of nature by the bishop of Paris, Étienne Tempier, in 1277, which will be discussed in the first chapter. According to Lindberg and Numbers, two of the most important promoters of an ecumenical vision of the relationship between science and religion, in addition to Brooke, this case illustrates the 'complexity of the

8

encounter between Christianity and science'. In particular, we are told that by losing some freedoms, philosophers of nature gained others. By accepting the idea of the omnipotence of God, scholars could now consider, for example, the existence of the vacuum, declared impossible by Aristotle. The authors are here summarizing the well-known thesis advanced at the beginning of the twentieth century by the physicist-chemist and Catholic philosopher Pierre Duhem, who went as far as to say that the condemnation of Aristotle's philosophy opened the door to modern physics.[20] This kind of reasoning is obviously spurious. Under the guise of the 'complexity' of events, looked at from the limited point of view of an abstract history of pure ideas, we are in fact asked to consider the positive side of censorship. According to this way of thinking, we should thank the Catholic Church for having sentenced Galileo to house arrest for the remainder of his life, because it has allowed him to write his *Discourses and Mathematical Demonstrations Relating to Two New Sciences*, published in 1638, five years after his condemnation by the Inquisition. As usual, analyses such as these reduce everything to individuals' beliefs and ideas and ignore institutions and the relative autonomy of science with regard to other social spheres.

The search for a middle ground also has the effect of minimizing the importance of the conflicts between science and religion that have occurred since the beginning of the seventeenth century. It is noteworthy that in the 350 pages of Brooke's book on the subject, a now classic reference in the field, there is but a single mention of the Index of Prohibited Books created by the Catholic Church, referring to the seventeenth century, and only to note that 'it is important not to exaggerate the oppressive effects of Index and Inquisition.'[21] After having read this survey that ends in the 1980s, the reader may thus legitimately think that there were no more scientific books put on the Index from the eighteenth to the twentieth centuries. But as we will see in chapter 4, this was far from the case. Another way to marginalize institutional conflicts is to frame the relationship between science and religion in purely intellectual terms. Thus, Brooke offers a cursory overview of Buffon's *Theory of the Earth* and simply notes in passing that his reconstruction of the earth's history, together with a hypothesis for its origins, 'exposed him to the charge of impiety from ecclesiastics and recklessness from other naturalists'.[22] In what is after all a book devoted to the relationship between science and religion, this vague formulation overlooks the fact that the Sorbonne's Faculty of Theology, in its role as official censor, had compiled a precise list of reprehensible propositions contained in Buffon's book, and that Buffon was forced to publish a retraction, as we will see in detail in chapter 4.

Similarly, in a book on the same subject that covers the period from Copernicus to Darwin, Richard Olson also presents Buffon's ideas noting only that 'to minimize his obvious divergence from Christian tradition, Buffon posited six epochs of earth history corresponding to the six "days" of creation, but each lasting thousands of years'.[23] Once again, nothing is said about the precise conflict with the theologians of the Sorbonne, probably because that would be to lay too much emphasis on 'conflicts' given the current trend towards ecumenism, 'dialogue', 'conversations' and 'encounters'. Finally, the same kind of subtle sophistry can also be used to simply deny the evidence. The historian Peter Harrison thus goes so far as to say that the condemnation of Galileo is not an example of a conflict between science and religion but instead reflects conflicts inside science and inside religion! As if debate between scientists and between clerics *excluded* the possibility of a conflict between science and religion. This fallacy also overlooks the institutional dimension of the condemnation of Galileo.[24]

A similar kind of sophism is provided by the title of a recent book suggesting that the idea that 'Galileo went to jail' is a myth. The argument is subtle. It consists in implicitly defining a 'prison' and 'imprisonment' as consisting *only* in being *in a cell of the Inquisition*. The proof is then easy since we know that Galileo was given permission to reside in the apartments of the Tuscan ambassador before finally residing in his house in Arcetri for the rest of his life (see chapter 1). So, 'Galileo was never held in prison'.[25] The problem with these analyses is that they curiously pass over in silence an interesting historical fact; namely that Galileo himself concludes many of his letters by saying 'Dalla mia carcere di Arcetri', that is literally: 'from my *prison* of Arcetri'.[26] The historian should conclude that from *his own point of view*, Galileo was indeed in a prison, for that word does mean, according to an Oxford dictionary: 'a building in which people are legally held as a punishment for a crime they have committed'.[27] Though one can indeed say he was 'under house arrest', that is not incompatible with his feeling of being in a 'prison'.

The question of the conflict between science and religion, recurrent since the first third of the nineteenth century (as will be seen in chapter 5), involves primarily religious *institutions* and not the *personal* beliefs of scientists, who as individuals could (and can still) reconcile their religious beliefs (private) and their scientific practice (public).[28] It is therefore of little import to know which scientists were or were not believers to understand how the scientific community became autonomous institutionally, methodologically and epistemologically by excluding theological and religious arguments. In a sense, the personal dimensions of scientific practice, which can be influenced by religious beliefs, are filtered through the collective rules of the scientific community. Any science has of course metaphysical assumptions. But these are not individual beliefs. They are institutionalized and should not be confused with religious beliefs by vague terms such as 'quasi-religious' or even 'metaphysico-religious'.[29] That Newton personally believed that uniform space constituted the 'sensorium Dei' does not imply that natural philosophers of his time had to accept this belief to use his physics. As will be seen, an increasingly clear separation between scientific discourse and religious beliefs is at work between the seventeenth and the nineteenth centuries in all scientific disciplines, beginning with astronomy (Copernicus) and physics (Galileo), followed by geology (Lyell) and natural history (Darwin) and ending with the origins of man (Darwin) and the history of religions (Renan). And it is not only an *institutional* separation, but also an *epistemological* division that explains why even Catholic scientists never directly invoked God to *explain* a given natural phenomenon. In other words, methodological and epistemological naturalism were part and parcel of the process of autonomization of the scientific field vis-à-vis other social spheres. Moreover, the conflict is clearly both institutional and epistemological. Many of the defenders of the autonomy of the sciences were themselves priests, pastors or Jesuits who nevertheless knew how to distinguish their personal faith and the autonomy of scientific knowledge.[30]

It goes without saying that the various conflicts that have occurred over the centuries each have their own local social and political contexts that have influenced the shape and chronology of events. But beyond these spatial and temporal variations, it is possible to identify several structural invariants, provided we go beyond the level of the individual, the year, if not the week – a micro-scale often used in the most extreme forms of the cultural history of science. Since we have set out to unravel the relative autonomization of science in relation to the religious institutions in the 'longue durée', it is useful to recall what the philosopher Pierre Duhem wrote in another context:[31]

> whoever casts a brief glance at the waves striking a beach does not see the tide mount; but under the superficial to-and-fro motion, another movement is produced, deeper, slower, imperceptible to the casual observer; it is a progressive movement continuing steadily in the same direction and by virtue of it the sea constantly rises.

This book describes the divorce between religions and the sciences, as well as the many conflicts that have marked this process from the seventeenth century to the present. We have chosen to approach the

11

subject from the point of view of science and not from the vantage point of theology or religious groups. It would take another book to describe the history of the theological uses of the natural sciences, uses that have most often taken the form of a more or less explicit concordism that seeks to interpret the meaning of the sacred books in a way that makes them consistent with the latest results of science. These 'natural theologies' invariably seek to provide a rational basis for faith, and their basic arguments have not changed in four centuries. They invoke the beauty, the layout and the order of nature to try to demonstrate the existence of a supernatural being. Only the examples vary: from the 'watchmaker' or the complexity of the eye, common examples from the seventeenth to the end of the nineteenth century, to the present 'flagellum' in biology and 'big bang' and 'chaos theory', 'fine tuning' and 'anthropic principle' in physics: all seek to suggest or even demonstrate the necessary existence of a creator.[32] This rationalist vision of religion with its source in Thomist philosophy has been the basis of Catholic theology up to Vatican II, and is opposed of course to the fideists who view faith as the province of intimate beliefs and feelings with no necessary relation to science.[33] We leave it to historians of theology to write the gripping history of the uses of science for purposes of apologetics and of the ways theologians have adapted their interpretations of the Holy Scriptures to scientific discoveries that they could no longer reject or condemn.[34] It is sometimes an amusing story, given the naivety of some theological interpretations of physics and modern cosmology.

Finally, the increasingly complete separation between scientific and religious institutions has not prevented numerous religious groups from continuing to exert pressure (social and political) to limit the freedom of scientific research. Indeed, the various religious groups who continue to believe in the literal meaning of texts that they consider 'sacred', because inspired – or even dictated – by God, will probably never cease to try to limit the freedom of research on any subject that calls into question all or part of these 'revelations'. Since the 1980s, these struggles have mainly concerned the theory of evolution, have a global scope and affect both the Christian and the Muslim world. They represent the most recent examples of attempts to reduce the autonomy of science in the name of beliefs that some organized groups seek to impose on all.[35]

As will be seen in chapter 7, the rise of fundamentalist religious sects and of aboriginal spiritualities and beliefs in 'traditional' medicine reminds us that the practice of science is neither natural nor universal. It rests, as Max Weber said, on *cultural* assumptions that, if they are not shared, generate a dialogue of the deaf and create incommensurable communities of thought. These questions raised about contemporary scientific thought also remind us that the relative autonomy of science is a recent conquest, and that it would be naive to think that it is irreversible. In conclusion, we should remember that it is necessary to take sides, and to live with the consequences of those individual and collective choices.

Acknowledgements

I would like to thank my colleagues and friends who read the original French manuscript and shared their comments and suggestions, thus improving the content and style of this book: Vincent Bontems, Jean Eisenstaedt, Robert Gagnon, Sylvie Duchesne, Gilles Janson, Mahdi Khelfaoui, Jérôme Lamy, Camille Limoges, Pierre Lucier, Wiktor Stoczkowski, Françoise Olivier-Utard and Jean-Philippe Warren. Thanks also to Jean Bernier of Éditions du Boréal who encouraged me to finish this book and fulfil a promise of several years standing. Special thanks to my long-time friend Peter Keating (who had translated my first book twenty five years ago!), for the English translation which I have revised and adapted here and there to take into account the fact that I am here addressing a different audience than the one aimed at in the French edition. Thanks to Sarah-Émilie Plante for helping to locate existing English translations of many of the cited texts. Special thanks to John Thompson from Polity for his interest in publishing a translation and to Ian Tuttle for effective copy-editing. Thanks also to my friend and colleague Ian Stewart who read the final English version and made useful suggestions of content and style, and to Stephen Snobelen who read part of the proofs and made useful suggestions. It should go without saying that I alone am responsible for the content – and the tone – of this book. Finally, I would like to thank Monique Dumont for the compilation of the index.

— 1 —

THE THEOLOGICAL LIMITS OF THE AUTONOMY OF SCIENCE

Lost in an immense forest at night, I have only a small light to find my way; come a stranger who tells me: 'my friend, blow your candle to better find your way'. This stranger is a theologian.

Diderot[1]

Modern science, institutionalized in the seventeenth century, gradually emancipated itself from philosophy – the original source of all reflection on the physical, psychological and social worlds – and gave birth, in the course of the nineteenth century, to an increasingly fragmented domain populated with specialized and technical scientific disciplines. During this process, 'philosophers of nature', generalists who were often self-taught and who cultivated several areas of knowledge at the same time, gave way to more specialized 'scientists' often dedicated to full-time research on a particular subject matter. The invention of the term *scientist* to replace *natural philosopher* by mathematician, philosopher, and Anglican clergyman William Whewell at the beginning of the 1830s, stands as a linguistic signpost of this social transformation.

This chapter describes the progressive autonomization of science in relation to other spheres of society and the extension of its specific mode of thought to the whole of the natural and social world. This process also generated conflicts with theologians, specialists in rational discourse on the gods, as the Greek word *Theologos*, first found in Plato, indicates. The Christian Church took up this tradition of Greek philosophy and applied it to the interpretation of a single God. Spokespersons for the monotheistic religions (Christian, Muslim and Jewish) are thus the guardians of a long-dominant interpretation of nature, wherein profane knowledge was subjugated to a divine

14

knowledge revealed in texts considered sacred (the Christian and Jewish Bibles, and the Koran). Of course, only scientists working in areas likely to contradict a religious view of the world struggled with these authorities. For it is never science in general that enters into conflict with a given religion, but only particular sciences when they raise an issue that is already the subject of a particular theological interpretation. It is therefore not surprising that physics and astronomy were the first sciences to become problematic within the Christian context since these disciplines deal with questions that confront Christian cosmology head-on.

The confrontation between faith and reason goes back to the roots of Christianity, the first debates having drawn philosophers from the Greek polytheistic tradition into opposition with theologians and the fathers of the Christian Church. The latter were obliged to justify the tenets of their 'cult' to the cultivated followers of the Ancients and argue that their religion was indeed a true 'philosophy' in the Greek sense of the word and that the proponents of this new philosophy had rational arguments that would demonstrate the existence of a single God as opposed to the polytheism maintained by the hitherto dominant elites.[2] It was not, however, until the rediscovery of Aristotle's works on nature in the Middle Ages that these conflicts took on a clear institutional form. The question of the relationship between faith and reason also arose in the Muslim world in the famous (posthumous) confrontation between al-Ghazali (1058–1111) – for whom philosophy must submit to theology – and the great jurist and Aristotle commentator, Ibn Rushd, latinized as Averroes (1126–1198) – who defended the autonomy of philosophy.[3] In the Christian world, the teaching of Aristotle's natural philosophy in the university Faculties of Arts at the beginning of the thirteenth century also generated a jurisdictional dispute with the Faculties of Theology, who saw philosophy as a propaedeutic discipline subject to their control.[4] Public confrontations between philosophy, the rational discourse on nature, and theology, the rational discourse on a revealed God, gained importance only after theology had become a *social institution* that sought to control and sanction any and all discourse calling into question the dogmas and the teachings of the Church.

These multiple and recurrent struggles between science and religion are not simple confrontations between individuals of varying degrees of stubbornness, as is too often suggested, but an expression of institutional power relations. At issue is the autonomy and freedom of science vis-à-vis the pretensions of the churches to rule over science through its theologians.

Aristotle: the first conflict of the faculties

In 1798, the German protestant philosopher, Immanuel Kant (1724–1804), then seventy-five years old, published his last book: *The Conflict of the Faculties*. Coming from the author of the famous three 'Critiques' (of pure reason, of practical reason and of the faculty of judgment), one might assume that the book dealt with the *intellectual faculties*, but it did not. The topic was the freedom of thought within university faculties. Kant called for nothing less than the absolute independence and complete freedom of thought and public speech for professors in the faculty of philosophy and, especially, the end of the guardianship of theology over philosophy. The Faculty of Philosophy, he says, 'must be free to examine in public and to evaluate with cold reason the source and content of this alleged basis of doctrine, unintimidated by the sacredness of the object which has supposedly been experienced and determined to bring this alleged feeling to concepts'.[5] Moreover, when 'the source of a sanctioned teaching is historical, then – no matter how highly it may be recommended as sacred to the unhesitating obedience of faith – the philosophy faculty is entitled and indeed obligated to investigate its origins with critical scrupulosity'.[6] Since philosophy sought truth, it must be free and 'subject only to laws given by reason, not by the government'.[7] Notice that, in Kant's time, philosophy included natural philosophy and therefore the sciences.

As a philosopher, Kant had himself suffered from the limits imposed on the freedom of expression when he had, four years earlier, published an essay on *Religion within the Bounds of Mere Reason*. He had then received an admonishing letter from the King of Prussia, Frederic William II, who accused him of misusing his 'philosophy to distort and disparage many of the cardinal and basic teachings of the Holy Scriptures and of Christianity'.[8] Kant had accepted his monarch's censorship and promised to no longer speak publicly about natural or revealed religion during the King's reign. Kant did not write again until after the King's death in 1797. As the new king, Frederic William III, had disposed of the theologian who dictated the religious policy of his father, Kant felt free to publicly expose the foundation of the problem: the guardianship of theology over philosophy within the universities. Kant remained the bugbear of Catholic theologians. As if to remind him posthumously of the limits of his critical discourse, his famous *Critique of Pure Reason*, even though published in 1781, was placed on the Index by the Catholic Church in 1827.

16

Eighty years later, with the Church in full anti-modernist crisis, Cardinal Mercier would even declare that Kant and Darwin were the sources of the 'modernism' that Catholics must resist.[9] Again in 2006, Pope Benedict XVI launched a similar charge describing the German philosopher as the theorist of the 'self-limitation of reason', a thesis that had subsequently been 'radicalized by the impact of the natural sciences' and should now be called into question.[10]

Theology's guardianship of philosophy in the Christian world, publicly denounced by Kant, dates back to the very origins of the medieval university. Indeed, Aristotle's writings on nature began to circulate at the beginning of the thirteenth century, and increasingly became a subject taught within the arts faculties of universities, particularly at the University of Paris.[11] Written four centuries before the birth of Jesus, the works of the Greek philosopher are of course 'pagan' and present an uncreated and eternal nature, in other words, a nature that is obviously incompatible with biblical and koranic stories of creation. The 'heretical' nature of the affirmation of the eternity of the world had already been noted in the Middle Ages by al-Ghazali in his denunciation of Aristotle's philosophy.[12]

As early as 1210, a provincial council of bishops in Paris had prohibited the public reading of Aristotle's works dealing with natural philosophy under penalty of excommunication. Reminders were issued in 1215 and in 1231 by Pope Gregory IX who demanded that the prohibited works be examined in order to expunge them 'of any suspicion of error', in order to retain only the useful parts.[13] According to the Pope, the secular sciences must serve 'the science of Holy Scriptures'. Accordingly, Christ's faithful were to study them only 'to the extent that it is proven that they condescend to the will of the sovereign master'. It was therefore legitimate, he said, to retain only the useful parts and to 'scrupulously' remove 'all errors liable to scandalize and offend readers'.[14] In this way, the beginning of the thirteenth century saw the establishment of a system of inquisition, the ecclesiastical censorship of books and the excommunication of delinquent authors whose dubious teachings were reported and denounced to religious authorities.[15] This system underwent consolidation in the sixteenth century in response to the Protestant Reformation and gave birth to the Holy Office (the Inquisition) in 1542 and to the Congregation of the Index in 1571.

Despite repeated warnings, the teaching of Aristotle continued to advance. The institutional battle between philosophers and theologians led to the promulgation of new statutes for the Faculty of Arts in 1272 and to the most famous condemnation promulgated in

March 1277 by the Bishop of Paris, Étienne Tempier, a condemnation quickly followed up in England by that of the Archbishop of Canterbury. The Tempier condemnation listed 219 propositions considered heretical, several of which were based on Aristotle's teachings, while that of his English colleague, Robert Kilwardby, included only thirty propositions.[16] The idea that the Faculty of Arts, responsible for the teaching of philosophy, must submit to theology was clearly expressed by Tempier. From the outset, he affirmed that he had learned from 'eminent and serious people, animated by an ardent zeal for the faith' that in Paris 'some men teaching the Art, *trespassing the limits of their own Faculty*, dare expose and dispute in their schools, as if it was possible to doubt their falsity, some obvious and execrable errors' – which he then went on to list. Table 1.1 presents a few that target Aristotelian natural philosophy.[17]

In response to this act of censorship, Godefroid de Fontaines

---

Table 1.1

Some of the propositions condemned by Bishop Tempier in Paris in 1277

4. Nothing is eternal on the side of the end that is not eternal on the side of the beginning.

9. That there was no first man, nor will there be a last; on the contrary, there always was and always will be repeated.

34. That the first cause could not make several worlds.

35. That without a proper agent, as a father and a man, a man could not be made by God [alone].

37. That nothing should be believed unless it is self-evident or could be asserted from things that are self-evident.

38. That God could not have made a prime matter without the mediation of a celestial body.

40. There is no more excellent status than doing philosophy.

49. That God could not move the heavens [that is, the world] with rectilinear motion; and the reason is that a vacuum would remain.

63. God cannot produce the effect of a secondary cause without the secondary cause itself.

90. That a natural philosopher ought to deny absolutely the newness [that is, the creation], of the world because he depends on natural causes and natural reasons. The faithful, however, can deny the eternity of the world because they depend upon supernatural causes.

(1250–1304), a philosophy professor in the Faculty of Arts of Paris, observed that:

> It is thanks to disputations, where one tries to defend one and the other of the positions involved, in order to find the truth, that the latter is best discovered. To hinder this method of investigation and establishment of truth is to obviously prevent the progress of those who study and seek to know the truth.[18]

The institutionalized disciplinary hierarchy that separated philosophy and theology, which made the first the handmaid of the second, was consolidated in the middle of the thirteenth century. It remained the basis of the conflicts of authority that arose in the course of the centuries that followed in the Christian world until the middle of the 1960s when the Vatican II Council finally abandoned the attempt to control the thoughts of the faithful. It was not until 1966, when Paul VI abolished the canons concerning banned books, that the Index officially ceased to exist.[19]

Despite Tempier's sally, the importance of Aristotle's works on physics and cosmology was such that they were finally accepted and reinterpreted in a manner consistent with the biblical teachings thanks to the work of philosophers and theologians of the thirteenth century. The most famous of these was the Dominican Thomas Aquinas, although initially some propositions central to his teaching also fell under the condemnation of 1277 for several decades.[20] Thus, after having been condemned, Aristotle's natural philosophy became, by the beginning of the fourteenth century, the official science in all the universities of the Christian world, where it provided the conceptual framework within which all observed phenomena gained meaning in agreement with the Bible and Church doctrine. In return for having accomplished the amazing feat of reconciling Aristotle and the Bible by transforming a pagan philosophy into a Christian philosophy – thereby resolving the first conflict of the faculties – Thomas Aquinas was canonized in 1323 and proclaimed 'Doctor of the Church' by Pius V in 1567. And at the end of the nineteenth century, Pope Leo XIII imposed Thomism as the official doctrine on all Catholic philosophers in his Encyclical *Aeterni Patris* of 1879, which sought 'the restoration of Christian philosophy according to the spirit of Saint Thomas'.[21]

In 1513, the Lateran Council adopted a decree that formalized the subordination of philosophy to theology, in such a way that, as Francesco Beretta explained, 'it is the light of revealed truth and not the fallacious science of this world that leads to true knowledge. The error of a proposition from a theological perspective entails its falsity: a proposal contrary to faith cannot be true, even from a strictly philosophical point of view. This principle justified the prosecution of those who supported such propositions and their condemnation as heretics'.[22] Christian orthodoxy concerning the relations between theology and natural philosophy, that is to say science, was undermined and questioned at the beginning of the seventeenth century by new discoveries and scientific theories. Still in the process of emergence, 'modern science' would become the site of a new confrontation whose climax was the condemnation of Galileo by the Inquisition in June 1633.

Copernicus tries his luck

While working on the draft of his great work, *On the Revolutions of the Celestial Spheres*, published in Latin in the Lutheran city of Nuremberg in 1543, the famous astronomer and Catholic canon Nicolaus Copernicus (1473–1543) was quite conscious that his proposal to place the Sun in the centre of the Universe and to make the Earth rotate on itself and around the Sun would upset the deep convictions of most of his readers – vey few as it turned out[23] – as well as all those who heard of the book through rumour and hearsay. From the very opening of the long letter addressed to Pope Paul III that served as the preface to his book, Copernicus admitted that he:

> reckons easily enough, Most Holy Father, that as soon as certain people learn that in these books of mine which I have written about the revolutions of the spheres of the world I attribute certain motions to the terrestrial globe, they will immediately shout to have me and my opinions hooted off the stage.

He further admitted having a 'long-continued hesitation and even resistance' to publishing his book but that his friends 'made [him] change [his] course'. Copernicus also reminded the Supreme Pontiff that 'books of mathematics are written for mathematicians'[24] and that 'if perchance there are certain "idle talkers" who take it upon themselves to pronounce judgment, although wholly ignorant of mathematics' and 'dare to reprehend and to attack [his] work' by 'shamelessly distorting the sense of some passage in Holy Writ to suit their purpose', he would take no heed and would 'even scorn their judgment as foolhardy'. And he added that it was 'in order that the unlearned as well as the learned might see that [he] was not seeking to flee from the judgment of any man' that he decided to dedicate his book to the highest representative of the Catholic Church 'rather than to anyone else'. And this, not only because of the dignity of his rank, but also for his 'love of letters and even of mathematics'. And above all, the authority of his judgment would 'easily provide a guard against the bites of slanderers, despite the proverb that there is no medicine for the bite of the sycophant'.[25] It is nonetheless ironic that the man Copernicus showered with so much flattery had just created the Inquisition (in 1542), a war machine dedicated to the pursuit of heretics that would ultimately condemn the astronomer's ideas in 1616.

Copernicus died on 24 May 1543, only a month after the publication of his book; we will therefore never know how he would have defended his ideas.[26] One thing is certain, *On the Revolutions of the Celestial Spheres* appeared in a politico-religious context dominated by the question of the Protestant Reformation.[27] The conditions of its publication were also hardly conducive to a positive reception by the Catholic Church: the book had been printed in Nuremberg, a centre for supporters of Luther, and by a printer considered a 'heretic' for having published Lutheran authors. In addition, Copernicus' first disciple, Georg Joachim Rheticus (1514–1574), who, in 1540, had published the *Narratio prima*, the first summary of Copernicus' theory, was himself a Lutheran and therefore a heretic. His name appeared in the Clementine Index of Prohibited Books of 1559 as a disciple of prohibited Protestant authors.[28] Well aware that this new theory could be interpreted as going against a literal reading of the Bible, Rheticus had even published a book trying to reconcile the Holy Scriptures and the movement of the Earth.[29]

Like many scholars of his time, Rheticus claimed the freedom to think about issues that concerned the natural world and not faith. It is very significant that the cover page of the *Narratio prima* quoted a sentence attributed to Alcinous, a platonic philosopher of the second century: 'the philosopher must be free in his judgment'.[30] This aphorism was often quoted, always in the original Greek form, by other promoters of the new sciences. It was cited by Johannes Kepler (1571–1630) in his *New Astronomy* of 1609 and by Galileo in his *Discourse on Floating Bodies* of 1612, as well as by Tommaso Campanella (1568–1639) in his *The Defense of Galileo*, drafted in 1616 – at the very time Galileo was denounced for his Copernicanism – but published only in 1622.[31] These quotations clearly suggest that behind the question of biblical interpretation lay the more fundamental issue of the autonomy of natural philosophy vis-à-vis theology.

Although published and promoted by Protestants, Copernicus' book did not create much controversy within the Catholic world in the years immediately following its publication, despite the fact that it was quickly denounced: first to the Holy Office by a Dominican Father – though, for unknown reasons, the Inquisition seems to have never followed up on this denunciation – and then by the philosopher and Lutheran reformer Philipp Melanchthon (1497–1560). Luther has also been said to have made critical remarks about Copernicus' system.[32] Unknown to Copernicus, a note 'To the Reader Concerning the Hypotheses of this Work' had been inserted, anonymously, at the beginning of his book by the Lutheran theologian Andreas Osiander (1498–1552), who was responsible for reading the final proofs before printing. Copernicus' ideas had circulated in abbreviated, manuscript form, as the famous *Commentariolus* that he had written at the beginning of 1510, well before the publication of his complete book. Thus Osiander's warning admitted at the outset that the novelty of Copernicus' hypotheses has already received 'a great deal of publicity' such that it left no doubt that 'certain of the savants have taken grave offense and think it wrong' to talk about the movement of the Earth. Osiander added, however, that 'the author of this work has done nothing which merits blame' because the task of the astronomer is not to find the truth since 'he cannot by any line of reasoning reach the true cause of these movements' but is, more simply, 'to provide calculations that fit the observations'. And to attain this objective, 'it is not necessary that these hypotheses should be true or even probable'.[33]

In sum, Osiander suggested that the Copernican model should be understood as a *mathematical hypothesis* useful for astronomical calculations, but it should not be assigned a *physical reality*, since the secular sciences could never attain the truth. The astronomical assumptions therefore did not have to be true, or even probable, and were only intended to produce valid calculations and predictions. This conception of astronomy, that can be described as pragmatic,[34] seemed useful because it allowed one to assert, in conformity with the Bible, that the Earth was in reality stationary at the centre of the universe, contrary to what Copernicus had said in his book. We also find this instrumental view of astronomy among Muslim scholars of the Middle Ages who sought to avoid conflict with the most conservative interpreters of Islamic doctrine.[35] In principle, this philosophical distinction between thesis (realistic) and hypothesis (simply pragmatic and descriptive) allowed astronomers to avoid any quarrel with theologians on the question of the truthfulness of the cosmology found in the Bible (or the Koran). As we will see later, this pragmatic epistemology was taken up by Cardinal Bellarmine in his opposition to the Copernican system. There is no doubt, however, that Copernicus believed in the reality of his model. Osiander's warning was, in truth, in flagrant contradiction with the preface that followed and that we have already cited, as well as with the entire content of the book itself, whose realist tone was as evident to readers of the time as it is for us today.[36]

Lutheran censorship of Kepler

As Copernican astronomers had to constantly justify the legitimacy of their discourse with regard to biblical assertions, Kepler had planned to respond to these objections in his first great book, the *Mysterium Cosmographicum* of 1596. He was then a young unknown astronomer and a mere teacher in a Lutheran school in Graz. His publisher had agreed to publish his work on the condition that the authorities at the Lutheran University of Tübingen offered some kind of endorsement. His former teacher, Michael Maestlin (1550–1631), who had introduced him to the Copernican system during his university studies, wrote a very positive report. In contrast, the Lutheran theologian Matthias Hafenreffer (1561–1619), also one of Kepler's former teachers and then rector of the university, was less enthusiastic. He suggested to his former pupil not to deal with the issue of the harmonization of the Scriptures with the Copernican system as that would only sow discord within his religious community. As Cardinal Bellarmine would later suggest to Galileo, he recommended that Kepler introduce the Copernican system only as an instrument of mathematical calculation. To ensure the publication of his book, Kepler acquiesced and limited himself to a brief statement at the beginning of the first chapter:

> Although it is proper to consider right from the start of this dissertation on Nature whether anything contrary to Holy Scripture is being said, nevertheless I judge that it is premature to enter into a dispute on that point now, before I am criticized. I promise generally that I shall say nothing which would be an affront to Holy Scripture, and that if Copernicus is convicted of anything along with me, I shall dismiss him as worthless. That has always been my intention, since I first made the acquaintance of Copernicus's *On the Revolutions*.[37]

This theological self-censorship led Kepler to suggest to his former teacher, Maestlin, who had informed him of the threats of their theologian colleague, a strategy of veiled speech: imitate the Pythagorean tradition of secrecy and keep silent in public; but if someone speaks to us in private, then we will offer them our frank opinion. To act otherwise would jeopardize the work and especially the livelihood of practitioners of astronomy, a discipline too complicated and too abstract for common mortals.[38] This tradition of secrecy was also a possibility envisioned by Copernicus in his dedicatory letter to the Pope where he wrote that he had wondered about 'whether it would not be better to follow the example of the Pythagoreans and certain others who used to hand down the mysteries of their philosophy not in writing but by word of mouth and only to their relatives and friends'.[39]

Kepler's institutional position became more secure during the next decade and made him more daring. When he succeeded Tycho Brahe as mathematician to the Emperor of the Holy Roman Empire, Rodolphe II, in 1601, he no longer needed the official support of his former university and felt authorized to publicize his views on the arguments against Copernicus derived from the Scriptures. He therefore devoted much of the preface to his New Astronomy, published in 1609, to refuting objections to the movement of the Earth, which were all based, he argued, on a distorted reading of sacred texts that overlooked the fact that such texts targeted readers who used the language of common sense and not the language of the learned. As for the authority of the theologians, Kepler considered it unnecessary to dwell upon it as it was simply not the responsibility of philosophy. While maintaining his respect for the 'Doctors of the Church', he recalled, ironically, the many past errors of the saints in the field of science and concluded that 'While in theology it is authority that carries the most weight, in philosophy it is reason'.[40] For both Kepler and his teacher Maestlin, astronomy was a discipline that had its own methods and that had nothing to do with theology and the Holy Scriptures. Theologians who attempted to control astronomers' discourse also annoyed Maestlin who considered them erudites who lacked a basic understanding of astronomy.[41]

Fortunately, the diversity of Protestant sects and the consequent lack of a centralized system of censorship meant that the heaviest sentence that could be applied was exclusion from the local community of practitioners, a fate that befell Kepler who was rather heretical in matters of theology. This decentralization did not entirely exclude the possibility of a more cruel punishment when secular control allowed it, as was the case in the Republic of Geneva, where, in 1553, the doctor and theologian Michel Servet had been condemned to be burned alive for 'heresy'.[42] As a response to the Protestant

Reformation, which led to the proliferation of Christian sects at the end of the sixteenth century, the Catholic Church consolidated its own institutions devoted to the monitoring and condemnation of speech and writings deemed contrary to its faith.

Galileo attacks theologians

It was not until the arrival of the Italian Catholic scholar Galileo Galilei (1564–1642) on the public scene that a major confrontation broke out over what interpretations should have precedence in questions concerning nature: those of the theologians, based on the Bible and the Fathers of the Church, or those of scholars, based on observation, calculation and reason? Galileo was convinced that the Copernican system described reality and was not simply a convenient means of calculation. Kepler thought likewise and had moreover discovered that the famous unsigned Preface to the *De Revolutionibus* had not been written by Copernicus, as many believed at the time, but by Osiander. Kepler revealed this fact publicly for the first time in the introduction to his 1609 *Astronomia Nova*.

Galileo's beliefs concerning the reality and the truth of the Copernican system had been in circulation in the Italian intellectual world since the publication of his *Sidereus nuncius* (*The Starry Messenger*) in 1610. He had announced therein the many discoveries made possible by his development of the astronomical telescope. These observations (the satellites of Jupiter, the mountains on the Moon and, later, the phases of Venus and sunspots) seriously undermined the dominant view of an immutable and perfect cosmos, and the book made him instantly famous throughout scholarly Europe.[43]

To defend himself against the growing number of devious and anonymous attacks that had begun in 1612, Galileo decided, at the beginning of 1615, to write a lengthy defence of Copernicus in the form of a letter addressed to the Grand Duchess of Tuscany, Christina of Lorraine. This letter, since become famous, developed an argument that he had first presented, in December 1613, in a shorter version, to his friend the Benedictine Benedetto Castelli (1578–1643), professor of mathematics at the University of Pisa, and the content of which had begun to circulate. A veritable treatise, this long letter defended the new astronomy against those who sought to 'spread among the common people' the idea that the proposals of Copernicus 'are against Holy Scriptures, and consequently damnable and heretical'.[44] Galileo reminded the reader, not without some exaggeration, that, once printed, Copernicus' book 'was accepted by the Holy Church, and it was read and studied all over the world without anyone ever having had the least scruple about its doctrine'.[45]

As the attack on the Copernican system had focused not only on physics or astronomy but also on issues of biblical hermeneutics, Galileo felt obliged to advance arguments on theological grounds that were in fact foreign to him. For although early theologians had already distinguished various levels of reading, including the allegorical, the dominant interpretation of the Bible remained the literal reading. Galileo thus developed in detail his own position on the relationships that should exist between the theologians' interpretation of the Sacred Scriptures and the interpretation of natural phenomena by philosophers. In doing so, he called into question the hierarchy of disciplines established in the Middle Ages and confirmed in the sixteenth century by the councils of Lateran and Trent. But Galileo was not alone in this thinking and Kepler, as we saw, had also stoutly defended the idea that one should not read the Bible in a literal manner and had distinguished the task of the theologian from that of the astronomer. But as Kepler was Lutheran, Galileo was careful never to refer to him by name.

Following Kepler, Galileo thus defended the idea that the new astronomy was not in opposition to the Holy Scriptures and therefore religion. Claiming to quote a high-ranking clergyman, he asserted that 'the intention of the Holy Spirit is to teach us how one goes to heaven and not how heaven goes'.[46] Just as Catholic as Copernicus, Galileo suggested one not confuse the genres and instead separate the 'purely physical propositions which are not matters of faith' from those 'supernatural propositions which are articles of faith'.[47] The interpretation of religious texts must, he insisted, always be adapted to the discoveries of empirical science and the sacred text could not be in error if it was correctly interpreted. On this point, he made frequent appeals through lengthy citations to Augustine (354–430), an uncontested authority. To show, for example, that the purpose of the Bible was not to answer scientific questions, he cited the Bishop of Hippo who had written that, on the question of the 'shape and arrangement of heaven' the Spirit of God 'did not want to teach men these things which are of no use to salvation'.[48] He also reminded his readers that Augustine had asked biblical interpreters to be cautious when it came to obscure topics and to avoid categorical assertions because one could then reject 'something which later may be truly shown not be in any way contrary to the holy books of either the Old or the New Testament'.[49] Galileo's letter to the Duchess emphasized the fact that, in his great work of astronomy, Copernicus 'never treats of matters pertaining to religion and faith, nor uses arguments dependent in any way on the authority of Holy Scripture, in which case he might have interpreted it incorrectly; instead, he always limits himself to physical conclusions pertaining to celestial motions and he treats of them with astronomical and geometrical demonstrations based above all on sensory experience and very accurate observations'.[50]

Galileo's view presupposed a certain degree of autonomy of philosophers of nature with regard to theologians at a time when, as we have seen, the latter ruled the roost. Galileo was well aware of this and admitted to the Duchess that he had 'some qualms' with respect to the pretensions of some theologians to enter debates about natural phenomena by claiming 'the right to force others by means of the authority of Scripture to follow the opinion they think is most in accordance with its statements', without, however, being themselves 'obliged to answer observations and reasons to the contrary'. He rejected the idea that 'theology is the queen of all the sciences and hence must not in any way lower herself to accommodate the principles of other less dignified discipline subordinated to her'.[51] It seemed to him obvious that no theologian having some knowledge of the other sciences would support the idea that 'geometry, astronomy, music, and medicine are treated more excellently and exactly in the sacred books than in Archimedes, Ptolemy, Boethius and Galen'. For this reason, he claimed that 'officials and experts of theology should not arrogate to themselves the authority to issue decrees in the professions they neither exercise nor study'. To do otherwise would be tantamount to behaving like an absolute prince who, 'knowing he had unlimited power to issue orders and compel obedience, but being neither a physician nor an architect, wanted to direct medical treatment and the construction of buildings, resulting in serious danger to the life of the unfortunate sick and in the obvious collapse of structures'.[52]

There was, however, a certain degree of naivety on Galileo's behalf in believing that the debate over the proper hierarchy of disciplines was purely rational. Although protected by his patron, the Grand Duke of Tuscany, Galileo underestimated the strength of those institutionalized hierarchical relationships that, since the middle of the thirteenth century, had placed natural philosophy in the service of theology. In such a system, science had to adjust its discourse in the light of theology and not the reverse. Galileo did his best to quote appropriate authorities, and religious friends helped in finding the citations from the Church Fathers that he used in his essay. The Spanish theologian Diego de Zúñiga (1536–1597), whom Galileo cited, had also, in his *Commentary on Job*, published in 1584 in Toledo and republished in Rome in 1591, stated that the movement of the Earth was not in contradiction with the Bible. The Carmelite Paolo Antonio Foscarini (1565–1616) thought the same and had also published a book on Copernicus and the movement of the Earth in 1615. But as will be seen later, both essays were put on the Index of Prohibited Books in 1616. Within the scholarly world, the deeply religious Kepler also claimed that theology should have no voice on issues concerning nature and that the movement of the Earth was not incompatible with a proper interpretation of the Bible.[53]

The Catholic epistemology of Cardinal Bellarmine

In 1615, when conflict between Galileo and religious authorities broke out, the Jesuit-trained Cardinal Robert Bellarmine (1542–1621) was a key player in the Roman Inquisition. Very influential in the Congregations of the Holy Office and the Index, he had a good knowledge of astronomy, having taught it at the Jesuit College of Louvain at the beginning of his career.[54] In the Vatican, Bellarmine embodied the hard line against the Reformation. Considered by his contemporaries as 'the hammer of the heretics', he was also renowned for having conducted the Inquisition trial against the philosopher Giordano Bruno (1548–1600), who, after seven years of interrogation, was burned at the stake for heresy in February 1600 in Rome on the Campo de' Fiori. Among his many heretical opinions, Bruno, a supporter of the Copernican system, claimed that the Universe was infinite and included multiple inhabited worlds. Bellarmine was also a noted theoretician of the relationship between Church and State. His importance within the Roman Curia was such that he was beatified in 1923, canonized in 1930 and declared a Doctor of the Church the following year, thus reaching the same status as Thomas Aquinas.[55]

Aware of the influence of the Cardinal, Foscarini had asked for his advice on his new book just as Galileo was writing his defence of Copernicus for the Duchess. The Cardinal sent him a brief response, because, as he put it, the Carmelite had 'little time for reading and [he] for writing'.[56] This missive is important because it sets out the epistemological framework for the interpretation of the Scriptures that long remained dominant within the Catholic Church.

From the outset, Bellarmine affirmed that Foscarini and Galileo 'are proceeding prudently by limiting [themselves] to speaking suppositionally and not absolutely as [he] had always believed that Copernicus spoke'. For if there was 'no danger' in dealing with the Copernican system as a simple useful hypothesis, to consider it as faithful to reality was in fact 'a very dangerous thing, likely not only to irritate all scholastic philosophers and theologians, but also to harm the Holy Faith by rendering Holy Scripture false'. He also noted that the Council of Trent 'prohibits interpretation of Scripture against the common consensus of the Holy Fathers'.[57] Finally, and most importantly, he formulated in advance a sort of response to Galileo, who, in his letter to the Grand Duchess of Lorraine, wondered how an opinion could be considered erroneous and heretical when it does not concern salvation.[58] Bellarmine told Foscarini that even on questions of cosmology, one cannot say that 'this is not a matter of faith since, if it is not a matter of faith "as regards the topic", it is a matter of faith "as regards the speaker"' because it is always the Holy Spirit who speaks 'through the mouths of the prophets and the apostles'.[59] Bellarmine thus reaffirmed Thomas Aquinas' stance laid out in his *Summa Theologica*, which generally represented the official position of the Church.[60]

In fact, Galileo knew Bellarmine's position perfectly well as his friend, Prince Federico Cesi (1585–1630) – founder in 1603 of the Accademia dei Lincei that had counted Galileo as a member since 1611 – had written at the beginning of January in 1615 to warn him that the Cardinal clearly considered Copernicus a heretic because 'the movement of the Earth is without doubt contrary to the Scripture'.[61] A month later, yet another friend, Giovanni Ciampoli (1590–1643), wrote to Galileo to inform him that, the night before, Cardinal Maffeo Barberini (1568–1644), a great admirer of Galileo and future Pope Urban VIII, had told him that it would be prudent to limit himself to the physical and mathematical domains for the theologians consider that the explanation of Scripture fall under their domain.[62] These theological positions echoed those of Bishop Tempier who, as we saw, warned philosophers at the University of Paris in his condemnation of 1277 to avoid issues considered the purview of theology.

At the epistemological level, the Catholic point of view can be summarized as having settled on the idea of the infinite power of God and the limited understanding of human beings. To believe, as Galileo did, that one could achieve certainty and understand the reality of nature amounted to playing God. Epistemological realism was therefore excluded and the only acceptable epistemological stance was a form of pragmatism or nominalism that sought to 'save appearances', that is to offer a plausible explanation of observed phenomena through hypothetical constructions that could not pretend to be true but only useful.[63]

The Inquisition and its procedures

At the time of Galileo's discoveries, which had aroused consternation among scholastic philosophers, theologians and Italian Catholic priests – who saw in them a questioning of the authority of the Bible – the institutions responsible for the control of thought were well established. Indeed, in response to the Reformation and to punish any criticism of Catholic dogma and thus reaffirm the authority and determination of the Roman Church, Pope Paul III, the very person to whom Copernicus had dedicated his *De Revolutionibus*, had formalized inquisitional practices by creating, in 1542, the Tribunal of the Roman Inquisition. The Church henceforth obliged all Catholics to report to local representatives of the Inquisition any fact or gesture that appeared dubious to faith or to the doctrine of the Church. The task of the members of the Holy Office was then to carry out an investigation by calling witnesses to confirm or disprove the allegations of the whistleblowers and then render a judgment. This machine of ideological control was supplemented by the institution of the Congregation of the Index in 1571 which was responsible for monitoring published books in order to better control, censor and prohibit the circulation of printed matter considered dangerous for the Catholic Church and faith. Following the invention of the printing press, these books multiplied during the sixteenth century, forcing the Catholic Church to create a new organization to control books.[64]

Novel in its Roman organization, the Inquisition had, however, a long tradition behind it. The Courts of Inquisition had been set up at the beginning of the thirteenth century by Pope Innocent III (1160–1216), and through the years and trials of heresy, inquisitorial practices had been codified and the procedures laid down. One of the most famous inquisitors of the Middle Ages, the Dominican Bernard Gui (1261–1331), had written an *Inquisitor's Manual* that explained how to proceed.[65] He thus bequeathed to his successors the experience he had accumulated as the Inquisitor of Toulouse. Later, in 1376, another Dominican, Nicolau Eymerich (1320–1399), also wrote a manual intended for bishops and Inquisitors, which was reprinted throughout the sixteenth century and 'updated' in 1578 in a new edition commented on by Francisco Peña (1540–1612), a doctor in canon law and a consultor to the Holy Office and the Congregation of the Index.[66]

Eymerich's *Manual*, together with Peña's additions and comments, explained in detail the procedures to be followed for the different types or degrees of heresy. There were, for example, heretical *propositions* and heretical *individuals*. The manual also identified seven criteria for the recognition of a heretic discourse that referred to the content of the Bible, the words of Christ transmitted by the Apostles, the assertions of the Church and the councils and to 'all that necessarily follows' from these principles. There were also eight criteria that qualified a person as a heretic, including doubting the faith or having an opinion that differed from the Church of Rome on an article of faith, having been excommunicated, opposing the Church or committing an error in the explanation of the Scriptures. And although the notion of error was broader than that of heresy, Peña concluded that, 'in matters of faith, error and heresy are synonymous'.[67]

The Inquisition and its courts therefore constituted a legal system with its own concepts and jurisdiction. Thus, an accused who stated that he was unaware of the heretical character of his opinions and who had 'always kept faith in his heart' would be 'compelled to recant as very suspect of heresy and assessed a hard penance'.[68] There was also a gradation in the suspicion of heresy. The *Manual* defined three types: weak, strong or vehement, and serious or violent. A proposition deemed merely 'reckless', while blameworthy, was subject to a simple warning as it was less serious than a statement considered suspect or 'vehemently suspect of heresy'. As we will see, it was this latter expression that was used against Galileo in 1632. Although the 'strongly suspect should not be considered heretics', they had nonetheless to 'abjure all heresies and in particular those of which they appear highly suspicious'. As for persons who, like Galileo, belonged in the category of penitents who 'recant only after being detained or simply cited one or more times by the inquisitor', they were to be 'treated with more rigor'.[69] Finally, as regards its jurisdiction, common blasphemers or 'those who are not opposed to the articles of faith' did not fall within the competence of the inquisitor. Similarly, diviners or ordinary seers did not concern the Inquisition.[70] In contrast, those who 'say for example that God cannot make the weather clear or make it rain, thereby directly oppose the dogma of the omnipotence of God' and thus fall under the thumb of the Inquisition.[71]

Galileo denounced, Copernicus sentenced

According to historian Francesco Beretta, the trial of the Roman Catholic Church against Galileo for his Copernican opinions began in 1615 and not, as is often held, in the fall of 1632 following the publication of his book on the two chief systems of the world (Ptolemaic and Copernican). Indeed, 'the trial of Galileo formally began on March 20, 1615, with the denunciation of the philosopher to the Tribunal of the inquisition by the Dominican Tommaso Caccini'.[72]

In fact, the gears of the Catholic bureaucracy began to grind earlier on 7 February 1615, when the Dominican Niccolò Lorini (1544–1617?) denounced Galileo's opinions in a confidential letter to the Prefect of the Congregation of the Index. He reported having 'come across a letter that is passing through everybody's hands, originating among those known as "Galileists" who, following the views of Copernicus, affirm that the Earth moves and the heavens stand still'. In his eyes, as in those of other members of his religious convent, the letter contained many propositions that seemed 'either suspect or rash' and he considered it his duty to alert the authorities to these facts.[73]

The text at issue was in fact a copy of a letter that Galileo had sent, in December 1613, to the mathematician Castelli, referred to above, in which he briefly presented his point of view on the movement of the Earth in relation to the Bible. He had been encouraged to put his thoughts on paper by the fact that the Duchess of Lorraine (mother of the Grand Duke of Tuscany, Cosimo II, for whom Galileo was the official mathematician and philosopher), had raised the question of the incompatibility of the ideas of Copernicus with the content of the Bible during a lunch at the palace in the presence of Castelli. The issue was topical because Ludovico delle Colombe (1565–1616), Aristotelian philosopher and enemy of Galileo, had published a book in 1611 in Florence entitled *Contro il moto della Terra* in which he noted, along with other arguments based on Aristotle's physics, the fact that such a movement was incompatible with the literal content of the Bible.[74] Galileo had seized this occasion 'to examine some general questions about the use of the Holy Scripture in disputes involving physical questions'.[75] Fortunately for Galileo, the expert report of the anonymous theologian asked by the Inquisition to assess the content of this letter concluded that there was nothing of importance to report and that apart from a sometimes improper use of certain words, its author 'does not diverge from the pathways of Catholic expression'.[76]

A month after Lorini's intervention, which did not lead to any accusation, another Dominican, Tommaso Caccini (1574–1648), appealed to the Tribunal of the Inquisition in Rome, at the Palace of the Holy Office, to denounce the fact that Galileo had publicly defended the Copernican doctrine on the movement of the Earth, which 'is contrary to the common account of almost all philosophers, all scholastic theologians and all the Holy Fathers'. No one, he added, is 'allowed to interpret divine Scripture in a way contrary to the sense in which all the Holy Fathers agree, since this was prohibited both by the Lateran Council under Leo X and by the Council of Trent'.[77]

In keeping with its procedures, the Inquisition convened two witnesses to confirm Caccini's charge the following November.[78] With the preliminary investigation completed, the theologians of the Tribunal of the Inquisition rendered their judgment on 24 February 1616 and concluded that the assertion that the Sun is at the centre of the world and stationary was 'foolish and absurd in philosophy and formally heretical insofar as it explicitly contradicts in many places the sense of Holy Scripture, according to the literal meaning of the words and according to the common interpretation and understanding of the Holy Fathers and the doctors of theology'. They also argued that the statement according to which the Earth was not the centre of the world and was not stationary was philosophically absurd but not strictly speaking heretical from a theological point of view, though it was 'at least erroneous in faith'.[79]

The next day, in light of this decision, Pope Paul V directed Cardinal Bellarmine to warn Galileo that he should abandon his Copernican opinions and that 'if he should refuse to obey, the Father Commissary, in the presence of a notary and witnesses, is to issue him an injunction to abstain completely from teaching or defending this doctrine and opinion or from discussing it; and further, if he should not acquiesce, he is to be imprisoned'.[80]

On 26 February, Galileo, who had been in Rome campaigning to avoid this harsh judgment, stood before Bellarmine and his witnesses as the Father Commissary General of the Holy Office told him that, if he did not obey this injunction 'the Holy Office would start proceedings against him'. The text concluded by noting that Galileo 'acquiesced in the injunction and promised to obey'.[81]

The procedure, which had begun with Caccini's denunciation, ended what may rightly be called 'the first trial of Galileo'. The investigation had indeed followed the procedures of the Inquisition and had led to a judgment to which Galileo was forced to comply under penalty of being declared 'relapse', in the inquisitors' words, a form of recidivism that automatically led to a strong condemnation. We will see further on that the famous judgment of 1633 against Galileo was based on this 1616 decision and would state that he had indeed broken his commitment by publishing a book promoting the Copernican system.

Alongside the Inquisition, the Congregation of the Index continued its work and published a decree on 5 March 1616 condemning books affirming the movement of the Earth. The introductory sentences of the decree clearly indicated the Congregation's role in censorship:

> In regards to several books containing various heresies and errors, to prevent the emergence of more serious harm throughout Christendom, the Holy Congregation of the Most Illustrious Lord Cardinals in charge of the Index have decided that they should be altogether condemned and prohibited, as indeed with the present decree it condemns and prohibits them [.]

And given the Church's pretension to universality, the decree specified that the condemnation of the books prevailed 'wherever and in whatever language they are printed or about to be printed'.[82] The books specifically targeted by the decree were those of Diego de Zúñiga and Copernicus, which were 'suspended until corrected', while that of the Carmelite Foscarini was 'completely prohibited and condemned'.[83] Its Neapolitan printer was even arrested.[84] In order to include any book on the same subject that might have gone unnoticed, the decree also prohibited, 'all other books' which teach the same doctrine. Despite this universal condemnation, which should have been sufficient, a decree explicitly condemning Kepler's *Epitome astronomiae copernicanae*, a synthesis of Copernicus' astronomy published in 1617, was adopted in 1619 and it was added to the list of prohibited books. Curiously, Kepler's previous and equally Copernican books, the *Mysterium Cosmographicum* and *Astronomia Nova*, were not mentioned. The Index officials probably considered that the books fell within the general prohibition of all works affirming the movement of the Earth, or that, as a Lutheran, the author was already heretical by definition.

All Catholics, 'of whatever station or condition' were therefore prohibited from printing, having printed, holding or in any way reading the proscribed works, and would otherwise be subjected to a 'penalty specified in the Holy Council of Trent and in the Index of prohibited books'.[85] The most serious penalty was of course excommunication. And, according to the rules of the Inquisition, any individual excommunicated for more than a year was regarded as heretical.

As soon as the decree became public, all those who possessed the books in question were expected to immediately hand them over to the local representatives of the Inquisition.[86] Of course, these prescriptions had no purchase in Protestant countries, where the incriminated works continued to circulate without censorship. Catholic scholars, on the other hand, had to seek written permission to read any prohibited book.[87] Thus, a disciple of Galileo who had become professor of mathematics at the University of Bologna wrote to Galileo in 1629 to tell him that he had yet to read Kepler's *Epitome* in which his astronomical theory was explained and that he had written to Mgr Ciampoli in Rome to ask for permission to read it.[88] Kepler's book had been put on the Index ten years earlier.

Perhaps expecting the worst, Galileo welcomed the fact that the decree of the Congregation of the Index in no way declared Copernicus' doctrine heretical, as some had claimed, but merely condemned those books having affirmed that there was no disagreement between the Bible and the doctrine of the movement of the Earth. The Cardinals of the Congregation thus maintained Osiander's original position, reiterated by Bellarmine, and adopted by a significant number of astronomers at the time.[89] Bellarmine had anticipated the likely outcome of this whole procedure from the outset. Galileo's friend who worked at the Vatican, Piero Dini (1570–1625) had written to Galileo at the beginning of March in 1615 to tell him that Bellarmine did not believe that Copernicus would be banned and that 'the worst that could happen to the book is to have a note added to the effect that its doctrine is put forth in order to save appearances, in the manner of those who have put forth epicycles but do not really believe in them, or something similar'.[90]

The official document specifying the list of corrections to the *De Revolutionibus* was not published until 1620. In the end, the cuts and changes affected only those few statements that affirmed the reality of his system too explicitly. Copernicus' letter to the Pope, cited above, in which he asserts that the doctrine did not seem to contradict the Scriptures, was also censored. This was, by the way, the very text that Galileo had quoted in his letter to Christina of Lorraine.[91]

Galileo claimed to be unconcerned by these judgments, considering that his 'behaviour in this affair has been such that a saint would not have handled it either with greater reverence or with greater zeal toward the Holy Church'.[92] Such a self-serving presentation might have contributed to the idea that the action taken by the Holy Office against him in 1615–16 was not a proper trial. Yet it is quite obvious that this ordeal had been generated by Galileo's attempt to have the reality of the Copernican system accepted. His promotion of the precedence of natural philosophers over the theologians in all questions concerning nature was intended precisely to avoid such a condemnation that ultimately also menaced his research programme insofar as he was forced to officially promise the inquisitors that he would never again speak about the movement of the Earth. He had moreover made the trip to Rome with the express purpose of trying to influence the decision of the Tribunal, but his behind-the-scenes actions had not influenced the final verdict. Moreover, the rumours that he had been formally condemned were such that prior to his return to Florence he had asked Cardinal Bellarmine to certify in writing that 'Galileo has not abjured in our hands, or in the hands of others here in Rome, or anywhere else that we know, any opinion or doctrine of his; nor has he received any penances, salutary or otherwise.' However, the Cardinal quickly added that 'he has only been notified of the declaration made by the Holy Father and published by the Sacred Congregation of the Index' that the doctrine of Copernicus 'is contrary to Holy Scripture and therefore cannot be defended or held'.[93]

Any knowledgeable observer understood that Galileo had suffered a major setback. This was evident in the comments of Julien de Médicis (1574–1636), then ambassador of Tuscany to Prague. In a letter to Paolo Gualdo (1553–1621), a friend of Galileo and a priest in Padua, written only three months after the proclamation of the Censorship, the ambassador claimed to have 'learned with great sorrow the storm that struck Lord Galileo' but admitted, ironically, to feeling 'that something good was derived from this evil, for those who enter some bushes can hardly get out without a few scratches'.[94]

In the months that followed the March 1616 prohibition of the promotion of the movement of the Earth, Francesco Ingoli (1578–1649), a priest viewed favourably by the Roman Curia and who was interested in astronomy, sent Galileo an essay in which he claimed to have demonstrated the impossibility of the movement of the Earth. Ingoli's arguments were quite traditional and were essentially the same as those of Tycho Brahe. A supporter of the immobility of the Earth, Brahe 'saved appearances' by proposing a mixed system in which the Sun continued to rotate around the Earth but wherein the other planets rotated around the Sun. Galileo considered this (third) system to be artificial and completely ignored it in his *Dialogue Concerning the Two Chief World Systems* published in 1632.

A faithful follower of the scholastic method, Ingoli enumerated twenty-two arguments of which thirteen concerned astronomy, five Aristotelan physics and four referred to theology. It is likely that the Roman Curia had commissioned this essay in order to publicly justify the decision of the Holy Office, which always kept the reasons for its judgments secret.

Never printed, Ingoli's letter nonetheless circulated widely. Galileo could not, however, reply without breaking the promise made to the inquisitors to no longer defend or publicly support the movement of the Earth.[95] Kepler, on the other hand, responded in 1618. In the conclusion of his long letter, he summarized his response to Ingoli's theological arguments saying:

> either one should not allow Christians to be astronomers [. . .] or, when given the freedom to seek the truth concerning heavenly phenomena, theology should reserve no right to restrict within narrow limits the search for truth or to impose laws from theology on astronomical science that is a totally different kind of science. [96]

Contrary to Galileo, who, in his letter to the Grand Duchess of Lorraine, had tried to find some consistency between the Copernican system and the Bible, Kepler pleaded for a complete separation of disciplines:

> Much as the astronomer wants to freely use his reasoning in astronomical matters without impediment on the part of theologians, so the astronomer should control himself so as not to infringe upon the rights of theology in matters of faith and morals and in order not to slide into heresy, by advancing reasons beyond the boundaries of his subject.[97]

Kepler had, moreover, criticized Galileo, without naming him, when he had commented on the fact that both Copernicus and his own most recent book, the *Epitome of Copernican Astronomy*, had been placed on the Index in 1619. Taking advantage of the publication of his *Harmony of the World* that same year, he inserted a letter addressed to Italian booksellers in which he first explained that he had written this book 'as a German, thus enjoying the freedom proper to the German custom, a freedom that is ever more greater through the confidence it puts into the loyalty of philosophers'. He then pointed out that his book 'can support and is not afraid of the usual censorship' practised in Italy. As to the censure of the Copernican system, he felt that it was only 'a consequence of the annual movement of the Earth around the Sun' and of the fact that some people 'mistakenly apply astronomical theories in an improper domain and use a method that is not appropriate to its object'. But, 'with due respect for the judgment' rendered by the Roman Church, he suggested to booksellers that they not sell his book to 'the vulgar'. As it was aimed at philosophers and reputable writers, booksellers should sell the book only 'to the most eminent theologians, the most renowned philosophers, the most exercised mathematicians and the most profound metaphysicians' who, as Copernicus' advocate, he could only access through librarians. Kepler even believed that after they had understood that this book, like his previous work, highlights 'the immense glory of Divine works', theologians would then come to understand and revise their previous judgment because they had not, at the time of their decision, access to the new evidence that Kepler now offered.[98]

We can surmise that Ingoli did not appreciate this way of separating astronomy from theology. In his final response to Kepler's letter, the priest reminded him that, among Catholics, theology was 'an architectonic science' and as such it also had 'the power to command the inferior sciences and consequently also astronomy so that the latter cannot go against the truth of the Scriptures'.[99]

In recognition of his philosophical and theological justifications of the decision of the Holy Office, Ingoli was immediately appointed consultor of the Congregation of the Index. In this capacity, he prepared reports on the works submitted to censorship and even contributed to the banning of Kepler's *Epitome of Copernican Astronomy* in 1619. Kepler, of course, was unaware of Ignoli's role in this matter. Ingoli also took charge of drafting the list of corrections to be made to Copernicus' book published in 1620 by the Congregation of the Index. This last effort marked the end of his role in this affair; Ingoli was subsequently appointed Secretary of the new Congregation for the Propagation of the Faith, established by Gregory XV in 1622 and which is known today as the Congregation for the Evangelization of Peoples.[100]

An 'admirable conjuncture' for Galileo

Following his crushing defeat, Galileo returned to physics and his scientific polemics, without, however, abandoning his convictions. Towards the end of the summer of 1623, the election of a new Pope created what Galileo considered an 'admirable conjuncture' likely to foster public discussion of the movement of the Earth. He even believed that if he did not take advantage of this exceptional situation, 'it was not conceivable', as far as he was concerned, 'to find another such opportune moment', for he was almost sixty and in poor health.[101]

The election of the new pope, Urban VIII, was, in effect, an unexpected pleasure for Galileo as Urban VIII was none other than the former cardinal Maffeo Barberini whom Galileo had always held in very high esteem. In 1620, Barberini had even composed an ode in Latin in Galileo's honour in order to express his admiration for his discoveries concerning Jupiter, Saturn and sunspots.[102] In addition, his very close nephew, Francesco Barberini, was devoted to Galileo, who had guided his studies in Pisa with Galileo's good friend Benedetto Castelli. Only two months before his election to the throne of Saint Peter, the cardinal had written to Galileo to thank him for the help he had given his nephew, who had just received his doctorate from the University of Pisa.[103] In another happy coincidence, the election occurred at the same time as the publication of Galileo's new book, *Il Saggiatore* (*The Assayer*), a pamphlet full of irony directed against the Jesuit Orazio Grassi who had criticized (with reason) Galileo's interpretation of the nature of comets. The book was printed under the auspices of the Accademia dei Lincei and dedicated to the new Pope.

Urban VIII had barely been inducted when, on 29 September 1623, Galileo began preparations to visit him in Rome to test the waters on the possibility of returning to his great work 'Discourse on the Tides'. This title, a perfect expression of Galileo's point of view, would ultimately be refused authorization for publication (*imprimatur*) in 1632 due to its physical and realistic and therefore insufficiently hypothetical character. Galileo was in fact convinced that the tides could be explained by the rotation of the Earth. In turn, this phenomenon stood as physical evidence of the Earth's movement. Moreover, he had written an essay on the subject in 1615–16 during the deliberations on the censorship of Copernicus, but had been unable to convince the cardinals to take it into account in their judgment.[104] Since then, his oath had forbidden him from addressing these issues.

Galileo went to Rome in the Spring of 1624. During his six-week stay, he saw Urban VIII on six occasions and received presents and the promise of a pension for his son Vincenzio.[105] The cardinals close to Galileo had discussed the question of Copernicus with the Pope, who had told them that his astronomical system had never been considered strictly speaking heretical but merely reckless. It could therefore be discussed without risk provided one kept to the astronomy, avoided discussions of theology and made an impartial presentation of all the arguments for or against the different systems.[106] In substance, the Pope did little more than reaffirm Bellarmine's position.

Galileo therefore returned to Florence convinced that he could finally discuss Copernicus in the open, and he returned to writing what would become his great work, the *Dialogue Concerning the Two*

Chief World Systems, namely those of Ptolemy and Copernicus. As we noted earlier, he completely ignored Tycho Brahe's mixed system that had found favour among the Jesuits as it saved appearances by leaving the Earth in the centre of the universe and the other planets rotating around the Sun (itself in orbit around the Earth as in the Ptolemaic system).[107] One can easily understand that, for Galileo, such a mixture was arbitrary and lacked what he considered the simplicity of the Copernican system.

Despite Galileo's optimism, the guardians of orthodoxy remained vigilant. In 1625, Galileo learned that the Inquisition had received a new denunciation made by a 'pious person', charging that *The Assayer* contained several statements in favour of Copernicus.[108] The charge was without foundation as the incriminated book had nothing to do with the issue and dealt instead with another delicate subject, atomism.[109] As will be seen in chapter 4, this complaint did not give rise to any charges, and Galileo saw this as a sign of support within the Roman Curia.[110] However, he was reminded in June 1624 that the cardinals who had discussed the Copernican question with the Pope had noted that since 'all heretics [Protestants] take Copernicus opinion as certain', he needed 'to be very prudent in his positions on this subject'.[111]

A trial balloon: the letter to Ingoli

As if to test the waters, Galileo decided to finally respond to Ingoli's essay against the movement of the Earth. By choosing to address Ingoli in 1624, Galileo clearly sought the attention of the upper reaches of the Roman Curia. But whereas Ingoli had dealt with the physical, astronomical and theological arguments against Copernicus, Galileo, following the counsel of his friends, avoided the theological questions, 'at least for the moment' and responded only to the first two types of arguments.[112]

Four times longer than Ignoli's essay, Galileo's response set out the reasoning which would later be taken up in his *Dialogue* – in which, it is worth remembering, he also avoided any theological discussion. He broached the subject cautiously and the letter to Ingoli constitutes in this respect a major about-face in relation to the content of his famous letter to Christina. Although it was never published, it was copied and circulated widely.

With regard to Ingoli's arguments against Copernicus, Galileo wrote that it was 'the opinion which [he] *then* considered true'.[113]

Justifying his silence on the theological arguments by noting that one must proceed differently with them as 'they are not subject to refutation but only liable to interpretations', he elaborated a sophistic argument that, even according to his friends, nobody would take seriously. He claimed that he did not 'undertake that task with the thought or aim of supporting as true a proposition that has already been declared suspect and repugnant to a doctrine higher than physical and astronomical disciplines in dignity and authority'. Rather, he wrote this letter 'in opposition to heretics, the most influential of whom [he] hear[s] accept Copernicus' opinion'. He thus claimed to show that if Catholics like him continue to be certain of 'the old truth taught [them] by the sacred authors', it was not through ignorance or lack of understanding, 'but rather because of the reverence [they] have toward the writings of [their] Fathers and because of [their] zeal in religion and faith'. The heretics may well blame him as a person who is 'steadfast in [his] beliefs, but not as blind to or ignorant of the human disciplines'. For, added Galileo, it should not bother a true Catholic Christian to be laughed at by a heretic 'because he gives priority to the reverence and trust which is due to the sacred authors over all arguments and observations of astronomers and philosophers put together'.[114]

It is difficult to read this text without suspecting that it is a sham. Indeed, Galileo went on this way for many pages, elaborating all the physical and astronomical arguments that showed beyond doubt the superiority of the Copernican system over those of Ptolemy and Tycho Brahe all the while no doubt assured that the reader would read between the lines and thus be convinced. He concluded his letter by saying that he hoped to treat the subject in more detail if he were to be given 'the time and strength to finish [his] Discourse on the Tides'.[115] In the end, his letter was never sent to Ingoli as Galileo's friends believed that this would be detrimental to his cause given that it would be obvious that no well-informed member of the Roman Curia could believe that Galileo had written what he truly thought. Thus, Mario Guiducci told Galileo, in April 1625, that in the letter to Ingoli 'Copernicus's opinion is explicitly defended and though it is clearly stated that this opinion is found false by means of a superior light, nevertheless those who are not too sincere will not believe that and will be up in arms again'.[116] In fact, even people of good faith would have had difficulty taking Galileo seriously given that many of these statements were contrary to what he had always maintained in private.

Galileo condemned and sequestered

In 1630, more than five years after having tested the waters with the letter to Ingoli, Galileo had largely completed his great work. He had transformed his *Discourse* into a *Dialogue* but had kept the reference to the tides in the title. He had only to seek the *imprimatur* of the city in which he hoped to publish. As he sought Rome's attention, he was obliged to negotiate with the Master of the Sacred Palace who was responsible for granting the license to print. The position was then occupied by the Dominican Niccolò Riccardi (1585–1639) whom Galileo considered an ally, as Riccardi had been the Consultor responsible for reviewing Galileo's previous book, *The Assayer*, in 1623. Riccardi had then written a very enthusiastic report to the Congregation of the Index, which had approved the publication. Riccardi had been appointed Master of the Sacred Palace in 1629, just as Galileo was finishing his book.

After many vicissitudes that are well known and need not be reiterated here, Galileo received the *imprimatur* from both Florence and Rome.[117] But the book was ultimately printed in Florence – a decision for which he would be criticized at his trial as the book was first supposed to be published in Rome. In the course of the negotiations with Riccardi (then responsible for censorship), however, Galileo was forced to abandon the original title of his book that referred to the tides as physical evidence of the rotation of the Earth and choose a more neutral title that referred only to the astronomical systems discussed in the book. He was also obliged not to take position and to explain clearly that physical proof of the truth of the Copernican system was not accessible to the human mind and that one therefore could always choose between the different systems to explain astronomical phenomena. Finally, he was forced to present the argument suggested by Urban VIII asserting that since God was all-powerful, one could not limit his power by claiming to know with certainty how he had built the universe.

Galileo accepted all this and, at the beginning of May 1631 – a year before the publication of the book – expressed to the Secretary of State of Florence his impatience with these numerous requests that had delayed publication, noting that the Inquisitor:

> could see with how much submission and reverence I agree to give the label of dreams, chimeras, misunderstandings, paralogisms, and conceit to all those reasons and arguments which the authorities regard as favouring opinions they hold to be untrue.

42

He went on to add that he sincerely 'never had any opinion or intention but that held by the holiest and most venerable Fathers and Doctors of the Holy Church'.[118] We too often forget that the very long title of the *Dialogue* on the astronomical systems of Copernicus and Ptolemy ends with 'propounding *inconclusively* the philosophical and physical reasons as much for one side as for the other'.[119]

The notice to the reader, which followed the dedication to his patron, the Grand Duke of Tuscany, used the language imposed by the Holy Office to grant the *imprimatur* and opened by recalling the condemnation of 1616, presented in terms that barely conceal, once again, Galileo's true thoughts:[120]

> Several years ago there was published in Rome *a salutary edict* which, in order to obviate to the dangerous tendencies of our present age, *imposed an opportune silence* upon the Pythagorean opinion that the Earth moves. There were those who imprudently asserted that this decree had its origin not in judicious inquiry, but in passion none too well informed. Complaints were to be heard that advisors who were totally unskilled at astronomical observations ought not to clip the wings of reflective intellects by means of rash prohibitions.
>
> Upon hearing such carping insolence, my zeal could not be contained. Being thoroughly informed about that *prudent determination*, I decided to appear openly on the theatre of the world as a witness of the sober truth.

And after having reminded the reader that he was in Rome in 1616 at the time of the Holy Office's decision, Galileo reiterated the arguments laid out in his letter to Ingoli and claimed (without laughing?) that his work aimed to, he proposed, 'show to foreign nations' that by collecting 'all the reflections that properly concern the Copernican system' he shall 'make it known that everything was brought before the attention of the Roman censorship' and that from Italy come 'not only dogmas for the welfare of the soul, but ingenious discoveries for the delight of the mind as well'.[121] He warned the reader that he approached the Copernicus system as a 'pure mathematical hypothesis' and that while he forcefully presented the evidence in its favour 'striving by every artifice to present it as superior to supposing the Earth motionless', it nonetheless remained that all this did not necessarily flow from 'any necessity imposed by nature'. He presented his theory of the causes of the tides not as evidence of the rotation of the Earth – of which he remained convinced – but only as 'an ingenious speculation' so as to be able to retain priority in what he considered to be his most important discovery. So, in order that 'no stranger may ever appear who, arming himself with our weapons, shall charge us with want of attention to such an important matter', Galileo 'thought it good to reveal those probabilities which might render this plausible, given that the Earth moves'.[122]

Galileo completed this appeal 'to the discerning reader', printed in characters that differed from those used in the remainder of the book – an element of style that will later be criticized at the trial as it suggested that Galileo was not the author of these words – by repeating that if the Italian scholars continued to assert 'that the Earth is motionless and holding the contrary to be a mere mathematical caprice', this was not from 'failing to take account of what others have thought' but rather 'for those reasons that are supplied by piety, religion, the knowledge of Divine Omnipotence and a consciousness of the limitations of the human mind'.[123]

Adhering to the conditions imposed by the Holy Office, Galileo had Simplicio, a character representing the supporters of Aristotle, utter the most important argument against the affirmation of the reality of the movement of the Earth, an argument that had been imposed by the Pope himself. While confessing that he did not entirely understand Galileo's reasoning on the causes of the ebb and flow of the sea and was therefore unable to judge them 'true and conclusive', Simplicio nonetheless raised in opposition 'a most solid doctrine that [he] once heard from a most eminent and learned person, and before which one must fall silent'. That doctrine affirmed divine omnipotence. Following this dogma, Simplicio concluded, it had to be admitted that 'it would be excessive boldness for anyone to limit and restrict the Divine power and wisdom to some particular fancy of his own'. This assertion ended the argument. Salviati, who spoke on behalf of Galileo, readily admitted that this was 'an admirable and angelic doctrine', that was 'well in accord with another one, also Divine' that directed the human mind to use its faculties to try to understand 'the constitution of the universe'. For even if one 'cannot discover the work of His hands', one can thus better recognize and admire the greatness of God and 'His infinite wisdom'.[124]

Despite all these desperate precautions used by Galileo to publish what he rightly considered as the work of his life, he was in the end naïve to believe that his enemies would be fooled by such cosmetics. Only a few months after the book was released in February 1632, the scandal broke out and Pope Urban VIII, feeling betrayed by Galileo, had one of those temper tantrums to which his entourage had become accustomed. At the end of July, the Pope ordered that the book be withdrawn from the market to be corrected if not completely prohibited.[125] Galileo's enemies began circulating the rumour that Simplicio was a simplistic character and therefore a little naïve if not actually an idiot and that Galileo had made him a spokesperson for the Pope. The choice of Simplicio, however, was a reference to one of the greatest commentators on the work of Aristotle, Simplicius, a sixth-century philosopher. He was in no way simple-minded but truly an expert in the scholastic natural philosophy then dominant. It was therefore quite natural that he represented the philosophical authority that advanced the Pope's argument.[126]

In order to confirm the orthodox character of the book, and before the Holy Office interfered and launched an inquisition, the Pope, who still considered Galileo a 'friend' – in the context of the court society of the Renaissance – undertook an exceptional procedure. He called for the establishment of a special commission of experts to carefully read the book and to tell him if Galileo had truly presented the system of Copernicus as the only system consistent with reality. The inquiry unearthed the 1616 document in which Galileo promised to no longer defend or promote the movement of the Earth. As the Master of the Sacred Palace, who held a generally favourable view of Galileo, confided to the Ambassador of the Grand Duke of Tuscany, 'this alone is enough to ruin him utterly'.[127] In effect, in light of that document, Galileo would be a 'relapse', which, according to the rules of the Inquisition, would result in a mandatory sentence of imprisonment and maybe even the pyre. The commissioners of the inquiry concluded that the case should be entrusted to the Inquisition, and the Pope ordered the local inquisitor of Florence to let Galileo know that he would have to appear before the Commissary General of the Holy Office in Rome. The importance of the case was such that the local Inquisition, which should have conducted the investigation for a book published in Florence, had been circumvented to the benefit of the Roman Inquisitor.

Deeply disturbed by his book's reception, Galileo, tired and in poor health – he was now nearly seventy – did everything possible to delay his trip to Rome, but, after months of negotiation, eventually gave up. The Tuscan Ambassador to Rome advised him to submit to the Inquisition and to retract 'in the way that its Cardinals will desire; otherwise you will encounter extreme difficulties in the solution of your cause'. In any event, he could not escape trial and a more or less severe punishment.[128] Galileo finally set out on 20 January 1633 and arrived in Rome on 13 February. Given his reputation, and out of respect for his patron, the Grand Duke, he was allowed to live, pending the start of the trial, in the Ambassador's apartment rather than in the prison of the Holy Office. Two months later, just before the official start of the trial, the Pope reminded the Ambassador that the topics discussed by Galileo were of 'great consequence for religion'.[129] In his report to the Grand Duke, the Ambassador, who had warned Galileo of the beginning of the trial, admitted to having found him 'extremely afflicted by this; and judging by how much I have seen him go down since yesterday, I have very serious worries about his life'.[130]

During the first interrogation on 12 April 1633, Galileo did not follow the Ambassador's advice and committed a strategic error in stating that far from defending the mobility of the Earth, his book showed 'the contrary of Copernicus' opinion' and 'that Copernicus' reasons are invalid and inconclusive'.[131] Surprised by this attitude, Vincenzo Maculano, the Commissary of the Holy Office in charge of the hearing and who wanted to quickly dispense with this delicate matter, spoke to the Pope via his nephew, Cardinal Francesco Barberini, another friend of Galileo. The accused had denied 'what is plainly evident from the book written by him' and did not seem to realize that as a consequence of this denial 'there would result the necessity for greater rigour of procedure and less regard for the other considerations belonging to this business'.[132] In less diplomatic and more transparent terms, the refusal to frankly admit his error would, according to the procedures of the Inquisition, lead to the use of torture. As indicated in Eymerich's manual of the Inquisitor, anyone who denied having 'intellectually adhered to a heresy' would be 'submitted to torture so that the Inquisitor could make his own opinion of the reality of the adhesion of the accused to true Faith'.[133] Worse still, 'the vehemently suspect of heresy, who would not abjure in front of the inquisitorial judge', would be burned at the stake.[134]

To avoid the possibility that the most respected scholar of Europe, a man shielded by the Grand Duke of Tuscany, be subjected to these procedures, Father Maculano obtained permission from the Inquisition to 'treat extrajudicially with Galileo in order to render him sensible of his error and bring him, if he recognizes it, to the confession of the same'. He therefore visited Galileo, and 'after many arguments and rejoinders had passed between [them], by God's grace', Maculano attained his objective and brought Galileo 'to a full sense of his error, so that he clearly recognized that he had erred and had gone too far in his book. And to all this he gave expression in words of much feeling, like one who experiences great consolation in the recognition of his error, and he would be willing to confess it judicially'.

The Inquisitor hoped that Galileo would stick to his promise, which would facilitate the conduct of the trial and maintain the reputation of the court, which could then 'deal leniently with the culprit'. Maculano concluded his letter by saying that if everything went well, the Tribunal 'might have [his] house assigned to him as a prison' as was suggested to him by Cardinal Berberini.[135] For there was no doubt that the final judgment was to be one of guilt with a severe penalty. Here again, the *Manual of the Inquisitor* is enlightening as to how Galileo was to be treated. In it, Eymerich explained that the penalty must carefully take into account 'the age, the level of instruction and the status' (layman, cleric, religious, etc.) of the culprit.[136] In his commentaries, Peña added that 'many inquisitors dealing with illustrious penitents do not put them in jail, but assign them to reside in a house or a castle'.[137]

The second interrogation by the ten Commissars of the Inquisition, all cardinals, on Saturday, 30 April, was limited to hearing Galileo's statement, two days after the informal meeting with Maculano. In his declaration, the aging natural philosopher affirmed that he had undertaken serious reflection since his first interrogation, some three weeks before, and had had time to reread his book. He admitted that the book appeared to have been written 'by another author' and that the style used did strongly argue in favour of Copernicus. He stressed in particular the arguments based on solar spots and the tides, which were in fact his favourite physical arguments as evidence of the reality of the rotation of the Sun and the Earth. Citing Cicero, he confessed to having been 'more desirous of glory than is suitable' and justified his error as the result of 'vain ambition'.[138]

Probably seeking to avoid the complete prohibition of his *Dialogue*, he added that in order to confirm most solemnly that he 'neither did hold nor do hold as true the condemned opinion on the Earth's motion and Sun's stability', he would be prepared to rewrite his book and add one or two days to his *Dialogue*, which then ended after four days of exchange between the protagonists. He would thus be able to 'reconsider the arguments already presented in favour of the said false and condemned opinion and to confute them in the most effective way that the blessed God will enable [him]'. He thus 'beg[ged] this Holy Tribunal to cooperate with [him] in this good resolution, by granting [him] the permission to put it into practice'.[139] Desperate, he must have told himself that his readers would once again decipher his true intentions.

Be that as it may, this curious offer was rejected, and after two further interrogations the Inquisitors rendered judgment on 22 June 1633. Having found Galileo 'vehemently suspected of heresy, namely of having held and believed a doctrine which is false and contrary to the divine and Holy Scripture', the Tribunal ordered 'that the *Dialogue* by Galileo Galilei be prohibited by public edict' and condemned its author 'to formal imprisonment in this Holy Office at our pleasure', the Inquisition reserved to itself 'the authority to moderate, change or condone wholly or in part the above-mentioned penalties and penances'.[140]

The verdict having been read before him, Galileo kneeled and 'having before [his] eyes and touching with [his] hands the Holy Gospels' recited the text of his abjuration. He admitted that he had been 'judicially instructed with injunction by the Holy Office to abandon completely the false opinion' of Copernicus and that he had nonetheless published a book treating of 'this already condemned doctrine and adduced very effective reasons in its favour, without refuting them in any way'. For this reason, Galileo affirmed that:

> desiring to remove from the minds of Your Eminences and every faithful Christian this vehement suspicion, rightly conceived against [him], with a sincere heart and unfeigned faith [he] abjure, curse, and detest the above-mentioned errors and heresies, and in general each and every other error, heresy, and sect contrary to the Holy Church; and [he] swear that in the future [he] will never again say or assert, orally or in writing, anything that might cause similar suspicion about [him].

He furthermore promised to denounce any heretic or anyone suspected of heresy to 'this Holy Office, or to the Inquisitor or Ordinary of the place where [he] happens to be'.[141]

Galileo's defeat and humiliation were now complete; the most conservative minds of the Roman Curia had overcome the moderates. As a further sign of the division that prevailed within the Church on this case, only seven of the ten cardinals signed the harsh judgment against Europe's greatest scientist. It is, furthermore, no doubt significant that the Pope's nephew, Cardinal Francesco Barberini, was absent during this humiliating meeting and did not sign the sentencing document.

So that other scholars might be made aware of their duty to obey, the Pope ordered that a copy of the sentence and abjuration be immediately sent to all the apostolic nuncios and Inquisitors and firstly to those of Padua and Bologna, so that all philosophy and mathematics teachers be properly informed.[142]

The next day, following jurisprudence, the prison sentence at the Holy Office was commuted to an 'assignation to residence' in his apartment at the Tuscan Embassy, where Galileo had remained throughout the trial. Six months later, the Pope authorized Galileo to move to his villa in Arcetri, near Florence, but only 'to live there in solitude, without summoning anyone, or without receiving for a conversation those who might come, and that for a period of time to be decided by His Holiness'.[143] Although the historian Peter Godman tells us that 'prison in perpetuity' meant, according to the customs of the Roman Inquisition, only three years in prison if the person was showing contrition, Galileo was never released from house arrest.[144] And as we noted in the introduction, he even concluded many of his letters with the mention 'from my prison in Arcetri'. He knew that his prosecutors would not make life easy. He confided to his friend Elia Diodati in July 1634 that the local Inquisitor had just told him that he should henceforth 'abstain from asking permission to return to Florence for otherwise he would return to Rome to the true prisons of the Holy Office'. Faced with such a response, the aging scholar added that he thought he could only 'draw the most likely conclusion that [he] would leave [his] prison only for another more common, narrower and long-term one'.[145]

Permission to stay in Florence was finally granted in February 1638, more than four years after his condemnation and after he had become completely blind, to allow him to be cared for. But he was unable to leave the house or walk in the city and especially forbidden to converse in 'open or secret' with anyone on the movement of the Earth under threat of 'the most serious penalty'.[146] Each request for a visit was subject to a decree by the Congregation. In April 1639, Galileo requested complete release, but his request was again rejected.[147] He thus remained sequestered until his death on 8 January 1642. He had, however, been allowed assistance by a young mathematician, protected by the Grand Duke, Vincenzo Viviani (1622–1703), who read him his correspondence, wrote letters dictated by his blind master and generally watched over him.

The fury of the attacks on Galileo can be measured by the fact that even his final testament was contested and at first considered invalid because, according to the inquisitorial law, heretics could not draw up a will and their property was supposed to be confiscated by the Church. Fortunately, the report of a consultor found the document valid because, having recanted, Galileo was not really a heretic; he had only been suspected of heresy, an entirely different matter.[148] As we will see in the next chapter, the Pope went so far as to refuse him a burial equal to his rank and reputation.

— 2 —

COPERNICUS AND GALILEO: THORNS IN THE SIDES OF POPES

[The refusal to liberate Galileo] might even one day be compared to the persecution of Socrates by his own country, a persecution that has been denounced by other countries and even by the descendants of his prosecutors.

Peiresc[1]

Because Galileo was considered one of the greatest natural philosophers in Europe during the first third of the seventeenth century, his conviction stood as one of the most painful episodes in the history of relations between the scholarly world and the Catholic Church. Even in the opinion of Pope John Paul II, it was the source of 'conflict and controversy and . . . misled many into thinking that faith and science are opposed'.[2] As we will see in this chapter, Copernicus and Galileo had long been thorns in the sides of the popes who succeeded to the throne of Saint Peter, thorns that many scientists tried in vain to remove. It took 200 years to expunge the works of these two great scientists from the Index of Prohibited Books and 350 years for Pope John Paul II to formulate an adequate response to the repeated requests of scholars from around the world seeking to rehabilitate Galileo and to bring the Catholic Church to admit that his conviction was an error.

Peiresc requests Galileo's release

While Galileo was alive, the astronomer and humanist Nicolas-Claude Fabri de Peiresc (1580–1637) probably fought the hardest for his release. At the centre of a vast network of correspondents in the

50

European Republic of letters, this councillor at the Parliament of Provence was held in high esteem by the politico-religious elite of his time.[3] He therefore took it upon himself, following a request by a friend of Galileo, to write to Cardinal Francesco Barberini – nephew of the Pope – in December 1634, to plead for the release of the 'poor Galileo'.[4] He asked the cardinal to 'excuse [his] boldness' in asking him to 'make some inquiries for the consolation of an old and sick septuagenarian, the memory of whom will be with difficulty erased in posterity'. He went so far as to eloquently predict that:

> future centuries will find it strange that after having retracted an opinion which had not yet been condemned in public, which was proposed only as hypotheses, such severity is deployed against an old septuagenarian man, taking him into prison or at least under arrest, so he could no longer go back to his city or receive visits and the consolations of his friends and be even deprived, because of distance, of the help and remedies made necessary by the infirmities and accidents to which he is subject. I say this with the compassion I have for Mr. Galileo Galilei, a good and elderly man, to whom I had wanted to write lately, only to learn from a friend in Florence that he was relegated to a country house, near a convent where his religious daughter, his only consolation, had died; and I was told that not only access to the city and his own house was forbidden, but that he was even not allowed to receive friends or write to them. This news broke my heart and forced me to shed bitter tears on the vicissitudes of human affairs: such evils, after having deserved such honour and glory that will last many centuries! I see that we forgave enormous and horrible crimes to excellent painters, and the most noble discoveries that have been made for centuries would not deserve indulgence for hypothetical opinions of an author who has never affirmed what we did not want to approve?
>
> Certainly such rigour would be found excessive in all countries, and even more by posterity than in the current century, where it seems that everyone forgets the interests of the public, especially those of the unfortunate, to think only of one's own interests. This affair will be a stain on this pontificate, if your eminence does not take it to heart and under his protection, as I beg and implore you humbly and with the utmost ardour, begging you to forgive me this freedom. But there must be a devoted servant who can sometimes give such marks of his attachment; because I do not think that those around you have the courage to manifest to you the thoughts they have in their heart and which are affecting the honour of your eminence.

The Cardinal answered politely but briefly that he would 'not fail to share with our Lord what you write to me concerning Mr. Galileo; but you will excuse me if I do not answer you with more detail on this point because, although the latest one, I am one of the cardinals attending the Holy Office'. Peiresc replied immediately in favour of his 'venerable old man' and reiterated to the cardinal that the refusal to show any leniency toward Galileo 'could even one day be compared to the persecution of Socrates by his own country, a persecution that has been denounced by other countries and even by the descendants of his prosecutors'.[5]

Galileo, who had been kept abreast of these intercessions, thanked Peirsec but confessed that he expected these actions to have little effect on what he considered a fortress that 'does not give any sign of yielding under the shock'.[6] Galileo was right. These supplications had no effect on Urban VIII's decision not to yield to pressures and Galileo remained under house arrest until his death in January 1642. As for Peiresc, he died in June 1637 at the age of fifty-seven.

Leibniz's defence of Copernicus

Although a Protestant, the philosopher Gottfried Leibniz (1646–1716) tried, towards the end of the 1680s, to convince the Catholic Church to annul the censorship of Copernicus. Conceding that the Church was infallible in matters of faith, Leibniz nonetheless felt it was wrong to force believers to adhere to theses that were unsustainable in terms of physics because these truths were not matters of will and could not therefore be changed at the discretion of the authorities.[7] In his exchanges with the Landgrave Ernst von Hessen-Rheinfels, a convert to Catholicism, he suggested that when the latter wrote to Rome:

> it would be useful to sound out Eminent Cardinals, if they would not be in the mood to lift the censorship published some time ago against the Copernican view of the movement of the Earth. For this hypothesis has now been confirmed by many reasons following from new discoveries, that the greatest astronomers have little doubt about it. Very skilled Jesuits (such as P. de Challes) have publicly admitted that it will be difficult to ever find another hypothesis that can account for all things so easily, so naturally and perfectly; and it is clear that nothing prevents him from promoting it openly, other than censorship. Mersenne, Minime, and Father Honoré Fabry, Jesuit, recognized and taught in their writings, that the censure was only temporary, until one better cleared things up, and it was deemed appropriate at that time only to obviate the scandal that this doctrine, then spread by Galilei, seemed to create in weak minds. We have now moved beyond that astonishment and every man of common sense easily recognized that

even if Copernicus' hypothesis would be true a thousand times, the Holy Scripture would not receive any damage. If Joshua had been a pupil of Aristarchus and Copernicus, he would not have stopped speaking as he did, otherwise he would have shocked his assistants as well as common sense. All Copernicans when they speak normally, and even among themselves when it is not about science, will always say that the Sun rose and set, and they will never say that of the Earth. These terms are assigned to phenomena and not the causes.

According to Leibniz, the Catholic Church should 'leave philosophers their reasonable freedom' (again echoing Alcinous) and he insisted that this 'censorship of Copernicus does harm since the most learned men of England, Holland and throughout the North (to say nothing of France), being almost convinced of the truth of this hypothesis, consider this censorship as an unfair slavery'. He even claimed that that it is to 'prostitute the Holy Scriptures and the Church to abuse their authority to prevent people from truths in matters of Philosophy'. He concluded his exhortation stressing that it 'would be possible to find some expedient, if one declared in Rome that all those who would like to argue that the hypothesis of Copernicus is true, should also declare that the Holy Scripture could not have spoken otherwise than it did and did not distance itself from the usual sense of the words'.[8]

We do not know if Leibniz's correspondent acted on his advice, but he himself attempted to put it into practice during his trip to Italy in 1689. But his efforts to persuade the Jesuits and other well-placed persons in Rome were to no avail. However, he didn't give up and in his *New Essays on Human Understanding*, written in 1703, but published in 1765, fifty years after his death in 1716, he lamented that 'they did not cease to continue in Italy and in Spain and even in the hereditary states of the Emperor to suppress the doctrine of Copernicus, to the great detriment of these nations whose scholars might have raised themselves to more beautiful discoveries had they enjoyed a reasonable and philosophic liberty'.[9]

The *Encyclopédie* appeals to Benedict XIV

The mathematician and physicist Jean d'Alembert, editor with his friend Diderot of the famous *Encyclopédie* named after them, took advantage of the entry on 'Copernicus' in Volume 3 of the *Dictionary of Sciences, Arts, and Trades,* that appeared in 1754, to suggest to Pope Benedict XIV (1675–1758) that he rectify the situation and cancel the prohibitions that had since become obsolete given that the Copernican system was now 'generally accepted in France and England'. He found it deplorable that 'the most enlightened philosophers and astronomers of Italy' do not dare to publicly support it or when they do they 'take great care to warn that they take it only as a hypothesis, and they are also very submitted to the decrees of the Supreme Pontiffs on this subject'. Using diplomacy and even flattery, d'Alembert added that it would be nice that 'a country as full of spirit and knowledge as Italy would finally recognize an error so detrimental to the progress of science'. For 'such a change would be worthy of the enlightened pontiff who governs the Church today; friend of science and scholar himself, it is to him to dictate to the Inquisitors the proper laws on these subjects'. Ironically, he added that perhaps one should, in these matters, follow the king of Spain who 'found it better to believe, on the question of the existence of the Antipodes, Christopher Columbus who came back from there, than Pope Zachary who never went there'. He rejoiced, in passing, that in France one finds himself 'much better to believe, on the world system, the astronomical observations than the decrees of the Inquisition'.[10]

The editor of the *Encyclopédie* clearly understood that Benedict XIV was himself embarrassed by his conservative wing in matters of modern science. Recently published, the first four volumes of the *Encyclopédie* had been reviewed by those in charge of the Index, and a Jesuit mathematician, rather favourable to Newton, had filed the report raising some points 'worthy of censure and blame' as indicated in the minutes of the meetings of the Congregation of the Index. But the Pope's influence managed to delay placing the famous *Encyclopédie* on the Index as it was only in 1759, one year after his death, that it was added to the long list of prohibited books.[11] Indeed, as soon as he was elected, Clement XIII, despite his name, proved himself less indulgent than his predecessor and immediately prohibited Catholics from reading the *Encyclopédie* under penalty of excommunication.[12]

We must, however, not confuse individuals and institutions. Conservative by nature, institutions are virtually forbidden to admit error without running the risk of losing credibility and authority. Thus, even though Benedict XIV admired the literary talents of Voltaire, who had even dedicated his play *Fanaticism, or Mahomet the Prophet* to the Pope, this admiration in no way prevented the immediate prohibition of the Italian version of the play and the condemnation of Voltaire's pamphlets by the Holy Office when deemed necessary.[13] For those who wanted to change practices within the Church, moving it toward greater freedom of speech in the area of physics, the only way forward was through patience, diplomacy and, above all, discretion, as any change had to be accomplished without attracting public attention.

Nonetheless, Benedict XIV's influence made itself felt early after his election. In 1741, the Congregation of the Index relaxed its stance somewhat and accepted the publication of a nearly complete edition of Galileo's works. The editor welcomed this event saying 'this very famous dialogue, so often clandestinely printed, finally appears in the light of day, equipped with all required authorizations and now offering itself for unrestricted public use'.[14] The edition, however, remained quite incomplete as it omitted the important letter to the Duchess Christina of Lorraine that had been the basis of Galileo's criticism of the literal reading of the Bible. In addition, permission to print was saddled with the obligation to also publish the sentence of the Inquisition as well as Galileo's abjuration. In the end, this 'freedom' was thus acquired at the price of modifications made to the original text. To transform Galileo's assertions into hypotheses consistent with the Decree of 1620, which made possible the reading of Copernicus' book after being censured of its most problematic sentences,[15] the publication's editor stated without shame that he had 'deleted or transposed in the hypothetical sense marginal notes that seemed not to go in that direction'.[16] As was the case for the text of Copernicus, these modifications were, in fact, in direct contradiction to Galileo's thought who never doubted for a moment that the movement of the Earth was real and not simply a useful hypothesis, even though no 'irrefutable' evidence had been provided. In both cases it was a comparative analysis of the plausibility of the systems of Ptolemy and Copernicus that, for Galileo, tipped the balance in favour of the latter. But, as we have seen, the rhetorical strategy of the defenders of the official position of the Church avoided such comparisons and simply asserted that there was no irrefutable proof to confirm the movement of the Earth, omitting of course to add that such irrefutable evidence to prove its immobility was also lacking.

The 'absolute proof and irrefutable fact' of the movement of the Earth only arrived at the end of the 1720s when the English astronomer James Bradley discovered the aberration of light from the stars, a phenomenon that could only be explained – in practice, of course, as absolute scepticism is always an option – by the movement of the Earth around the Sun. But only sceptics of the calibre of Cardinal Bellarmine could really demand such a 'proof' before abandoning the idea of an immobile Earth, given the theoretical coherence of the whole of physics that had been in place since Kepler's amendments to the Copernican system. With respect to the daily rotation of the Earth, there could not have been convincing empirical proof until Foucault's pendulum experiment in 1851. And to appreciate the epistemological complexity of scientific statements, one should recall here the great mathematician and philosopher of science Henri Poincaré's statement – then perceived as scandalous – at the beginning of the twentieth century according to which even the famous Foucault pendulum experiment did not really 'prove' that the Earth rotates.[17] For the nominalist Poincaré, it only constituted additional and convincing evidence but was not really the 'absolute' proof that Bellarmine always wanted, on the basis of an epistemology that greatly differed from that of many scientists of his time, including Galileo.

Galileo's late burial

At the time of Galileo's death on 8 January 1642, Urban VIII had still not forgiven his impudence in publishing the *Dialogue*. He had made it illegal for the Grand Duke of Tuscany to organize a lavish ceremony, to erect a mausoleum and to lay the body of the great philosopher of nature to rest in the Florentine church of Santa Croce. It would in fact have been unthinkable that a heretic like Galileo would have found official refuge in a Catholic Church. The pope had therefore made it clear to the Ambassador of the Grand Duke that it was not 'at all a good example to the world' because Galileo had been called before the Holy Office for having held 'a very false and very erroneous opinion, with which he had impressed many others and had given such universal scandal with a doctrine that was condemned'.[18]

Galileo's admirers did not stand down and chief among them was Galileo's last assistant, Vincenzo Viviani. A recognized scholar, Viviani had become a mathematician of the Court of the Grand Duke of Tuscany in 1666. Faced with the refusal of the authorities to recognize his master, he had himself erected a monument in Galileo's honour at the entrance to his own home in 1693. Despite its private character, in 1698 this monument was mentioned in an important Florence tourist guide that applauded Viviani's initiative and called for a formal recognition of Galileo as a national glory.[19]

It would in fact be necessary to wait for more than a century after Galileo's condemnation for the Church to finally accept that Italy might give its greatest scientist a burial worthy of his reputation. Even then, the decree of 16 June 1734 stipulated that the inscription for

Galileo's tomb must first be submitted to the Holy Office.[20] Finally, in March 1737, Galileo's remains were laid to rest in the Basilica of Santa Croce, facing those of Michelangelo (1475–1564) and Machiavelli (1469–1527), two other Tuscan giants. The event was political and expressed the desire of the Tuscan state for autonomy vis-à-vis the Holy See. It is also significant that, according to historian Paolo Galluzzi, no representative of the Church attended the ceremony.[21]

The end of harassment

In 1753, Benedict XIV succeeded in modifying the procedures used in the censorship of books so that the choice of the consultors had to be approved by the Pope himself. As a shining example of the glacial pace of change in the Roman Curia, he confided to a friend that the project, initiated in the second year of his pontificate, had taken him eleven years to complete.[22] This reform, which can in fact be interpreted as a further bureaucratization of the Congregation of the Index,[23] led naturally to the publication of a new edition of the Index of Prohibited Books in 1757. The new edition omitted, without comment, the phrase that condemned 'all other works' that affirmed the movement of the Earth, a phrase that was part of the original Decree of 1616 prohibiting support for the mobility of the Earth and the immobility of the Sun. It might be considered somewhat curious that the books specifically condemned on the basis of the Decree of 1616 (Copernicus, Galileo and Kepler) were not consequently removed from the new list, but this contradiction illustrates the Church's strategy of incremental change undertaken so as not to attract attention to actions that might be interpreted as an admission of error.

The new Index of 1757 did not go completely unnoticed, however. The French astronomer Jérôme Lalande, while passing through Rome in 1765, openly asked why Galileo's *Dialogue* had not been removed from the list. The fact that he failed to mention the names of the other scientists put on the Index shows that as far as he was concerned, even if Copernicus and Kepler were also great scholars, Galileo alone had become the symbol of science. According to Lalande's own account, the Prefect of the Congregation of the Index responded to his query by saying that since the Inquisition had condemned Galileo, this condemnation would have to be changed, which, as one might easily imagine, would require a long procedure leading to an uncertain conclusion.[24] As we will see in chapter 4, Lalande's own book popularizing astronomy, *L'Astronomie des dames*, would be put on the Index in 1830.

At the beginning of the nineteenth century, Newton's physics and his Copernican cosmology had been generally adopted throughout the scholarly world. While a number of books were printed with the more or less official approval of the Church, the more conservative minds of the Holy Office still sought to maintain the coherence of doctrinal decisions, despite their obvious absurdity.[25] It would take the crisis generated in 1820 by the application for permission to print a book on physics by an Italian canon, Giuseppe Settele, professor of Astronomy at the University La Sapienza in Rome, for the Holy See to finally put an end to the prohibition on teaching the Copernican system. Even though the consultor responsible for reading the book concluded that there was no longer any reason to oppose defending the movement of the Earth, the Master of the Sacred Palace, viewing the event from the lofty vantage point of principles, refused to issue the *imprimatur* arguing that the Decree of 1616, never repealed, forbade the defence of such a heretical idea. Somewhat comically, the stubborn refusal of the Master of the Sacred Palace to print Settele's book led the Cardinals of the Holy Office to issue a decree in 1822, approved by Pope Pius VII, that reminded the Master that 'there must not be any denial, by the present or by future Masters of the Sacred Apostolic Palace, of permission to print and to publish works which treat the mobility of the Earth and the immobility of the Sun, according to the common opinion of modern astronomers'. And the Decree added that 'those who would show themselves to be reluctant or would disobey, should be forced under punishments' left to the good judgment of the Holy Office.[26]

A logical consequence, but slow to happen, of this decision, was that the books of astronomy long prohibited by name (including those of Copernicus, Galileo and Kepler) were finally removed from the 1835 edition of the Index of Prohibited Books. So as not to attract attention, these withdrawals were, of course, made without any publicity. It may be wondered if the very stubbornness of the Master of the Sacred Palace was not intended to simply force the institution to say clearly what it had refused to say and to bring the decrees into conformity with the reality of the time. One thing is certain: Galileo was once again right. As he had ironically noted on a slip of paper found in one of his copies of the *Dialogue*: 'Beware theologians! Declaring as object of faith proposals that concern the movement and the rest of the Sun and the Earth, you expose yourself to the danger of having perhaps, with time, to condemn as heretics those who would contend that the Earth is stationary and that it is the Sun that moves'.[27]

Napoleon grabs the archives of Galileo's trial

The slow and tortuous acceptance by Catholic Church authorities of the scientific validity of the Copernican astronomical system, nearly three centuries after its publication, did not put an end to the criticism of those who demanded nothing less than a full recognition of the error that lay at the basis of Galileo's condemnation by the Roman Inquisition.

The symbol that Galileo's trial had become to rationalist minds (not to speak of those with anticlerical views) took on a dramatic new turn with the conquest of Italy by Napoleon. As one who loved to surround himself with scholars and who had brought with him the most prestigious of them during his expedition to Egypt (1798–1801), Napoleon ordered, in December 1809, that all the Vatican archives be repatriated to Paris.[28] Throughout the following year, thousands of folders moved at great expense from Rome to Paris including the precious documents surrounding Galileo's trial that France sought to publish in full. But the political instability of the following decades – the fall of Napoleon, the Restoration, and the July Monarchy – conspired to ensure that the folders, after many complaints from the Vatican, eventually returned to Rome in 1843 without ever having seen the light of day.[29]

The pressure to publish these documents increased several years later when, taking advantage of the instability engendered by the revolutionary Italian movements of 1848, Giacomo Manzoni, Minister of Finance of the ephemeral Roman Republic, and his friend, the physicist Silvestro Gherardi, Minister of Public Instruction, gained access to the Holy Office archives and quickly copied several documents from the Galileo file. But as the pontifical sovereignty of Rome was quickly restored, the Holy Office, fearing an 'ideological' use of the copied documents, decided to take the lead. In 1850, the prefect of the Vatican Secret Archives, Bishop Marini, who had contacted Paris as early as 1815 to reclaim the archives, published (in Italian) *Galileo and the Inquisition*, in which selected documents and extracts from the trial were made public for the first time. As the historian Annibale Fantoli has noted, the primary objective of this publication was 'to offer an apology for the behaviour of the Holy Office with respect to Galileo, by dissipating the suspicion of an inhuman rigor in the treatment of him during the trial of 1633'.[30] At the time, the (false) idea that Galileo had been tortured circulated widely among anticlerical groups. The publication did, however, resuscitate discussion of the whole affair. Archivists and historians debated the quality of the analyses and especially of the document transcripts, as the published dossier was far from complete.[31]

In the decades that followed, a number of books on the same subject appeared in English, French, Italian and German.[32] Henri de l'Épinois, archivist and French Catholic historian, was quite critical of Marini's work, whose cited documents, he suggested, were not satisfactory and even led one to believe that the Church was reluctant to let the facts be known.[33] He obtained permission from the Vatican to review the archival documents and in 1867 published *Galileo, His Trial, His Condemnation Based on Unpublished Documents*. Three years later, the Italian Silvestro Gherardi finally published the documents that he had consulted in the 1840s under the catchy title *The Trial of Galileo, Reviewed, Through Documents From A New Source*.

This resurgence of discussion surrounding Galileo culminated in the promulgation of a royal decree in 1887 announcing the project to publish a national edition of Galileo's complete works including his correspondence, all at the expense of the Italian Government and under the direction of the historian Antonio Favaro.[34] Bolstered by this state support, Favaro obtained permission to consult the archives of the Holy Office and of the Vatican to collect all official documents relating to the trial of the great Italian scientist. Composed of twenty volumes published between 1890 and 1909, this publication replaced the many more or less reliable previous editions. It remains the essential reference on Galileo.[35]

The rehabilitation of Galileo

The publication of documents is one thing, but Galileo's rehabilitation was another. Anniversaries always offer ideal opportunities to return to highly symbolic historical events.[36] In preparation for the 300th anniversary of the death of Galileo, the Pontifical Academy of Sciences, established in 1936 by Pius XII to better the relations between the Christian faith and modern science, undertook the initiative to publish a biography of the learned Florentine.[37] The underlying intention to defend the Church was explicitly laid out by the President of the Pontifical Academy who indicated that this biography would be nothing less than an 'effective demonstration that the Church did not persecute Galileo but it abundantly helped him in his studies'.[38]

By entrusting the project to the priest and historian Pio Paschini, the President was probably not expecting that Paschini might take a more critical view of history. When he submitted his manuscript for publication at the beginning of 1945, three years after the 300th anniversary, Paschini was surprised to see it refused by the Pontifical Academy of Sciences, which found it too partisan in Galileo's favour! Confronted by Paschini's refusal to amend his analysis, which he himself considered impartial and rigorous, the Academy referred the problem to the Secretariat of State of the Vatican, who submitted it to the judgment of the Holy Office. The latter confirmed, in turn, that the manuscript was indeed too favourable to Galileo and too critical of the Jesuits and the Dominicans, the two Religious Orders involved in the trial of 1633. Pius XII, who had nonetheless been favourable to the biographical project, supported his bureaucracy's judgment. The book was therefore censored, confirming once again the power of the more conservative minds over the Vatican bureaucracy.

Archbishop Giovanni Battista Montini, future Pope Paul VI, then Assistant to the Secretary of State and sympathetic to Paschini's point of view, was tasked with personally announcing the bad news to the unhappy author. Paschini responded that he was disappointed and dissatisfied with the decision that he considered unfair as he felt he had been impartial and had not defended Galileo.[39] Well aware of the divisions within the Church, he confided to a friend that he was firmly convinced 'that in reality the Holy Office from the beginning did not want at all such a publication. The Pontifical Academy of Sciences did want it; the Holy father approved it, but the Holy Office did not; the latter was only too happy to find a pretext for letting the matter fall by the wayside'. Paschini explained the decision by the fact that the 'superiors are always right, especially when they are wrong'.[40] He obediently submitted himself to self-censorship and was content to publish, in 1950, a short descriptive article on Galileo in *The Catholic Encyclopedia*, of which he was the editor-in-chief.

Galileo and Vatican II: a new admirable conjuncture

That the case of Galileo was a real thorn in the sides of Popes caught up in the ideological struggle between conservative and liberal factions within the Roman Curia, is even more obvious when we learn of the unexpected result of this curious history of censorship of a simple biography of a seventeenth-century scholar, commissioned by an organization (the Pontifical Academy of Sciences) controlled by the Vatican.

Deceased in December 1962, Paschini had bequeathed his manuscript to his former student Michele Maccarrone, also a priest and Paschini's successor to the Chair of Religious History at the Pontifical Lateran University in Rome, who felt compelled to publish the document. Maccarrone was then President of the Pontifical Commission of Historical Sciences and would become an expert at the Vatican II Council.[41] Recall that this ecumenical council had been reactivated by Paul VI in the autumn of 1963, following the death in June the same year of its initiator John XXIII. Convened to rethink the place of the Church in the modern world, Vatican II deeply marked the history of the Catholic Church.

Maccarrone thus returned to the Pontifical Academy of Sciences, which then proved itself more open to publication in the context of the 400th anniversary of the birth of Galileo (born in 1564). Just as Galileo had seen an 'admirable conjuncture' in the election of Pope Urban VIII, 1963 would provide a date with destiny for Paschini's two volumes of *Galileo Galilei*. As it turned out, Mgr Montini, who, as we saw, had been rather favourable to Paschini in the tug of war between the latter and the Holy Office, was elected Pope on 21 June 1963. During a hearing with the new Pope Paul VI, Maccarrone informed him of his project to publish Paschini's book and he received the support of the Holy Father. In addition, the debates of the Ecumenical Council that autumn favoured publication of the manuscript refused twenty years earlier. In effect, one of the issues debated by Vatican II was that of the relationship between science and religion. In this context, several members of the Council argued that an explicit statement on Galileo would be worthwhile given both his symbolic status and the unique opportunity provided by the 400th anniversary of his birth. They thus echoed the many Catholic scientists clamouring for a solemn rehabilitation of Galileo.

A preliminary version of what would become part of paragraph 36 of the Pastoral Constitution *Gaudium et Spes* on the 'legitimate autonomy of human culture and especially of the sciences' suggested the explicit mention of the fact that the conviction of Galileo was a mistake. However, the majority of the members of the Council refused to do so, and the resulting compromise left the text vague. However, a note at the bottom of the page referred, without comment, to the work of Paschini on Galileo that the authorities had finally agreed to publish, although not without having changed the more critical parts.

Maccarrone had not been made aware of these 'touch-ups', described as minor by the Jesuit Edmond Lamalle who was responsible for the final edition of Paschini's manuscript. In fact, under the pretext of updating a work completed twenty years earlier, these corrections clearly reversed several of Paschini's conclusions.[42] A Member of the Holy Office representing the conservative current, Mgr Pietro Parente, felt the book offered nothing new. Also a participant in the Second Vatican Council, he was strongly opposed, along with others, to any frank admission that the trial of Galileo was a mistake.[43]

The refusal by some of the archbishops to discuss Galileo might also have had a cultural foundation. During the discussions held on 11 February 1965, Gabriel-Marie Garonne, Archbishop of Toulouse, cautioned that if the case of Galileo was not mentioned, the scientific community would be disappointed. Mgr Zoa, Archbishop of Yaoundé, Cameroon, replied that one must distinguish between the local Church and the universal Church and that the case of Galileo was of particular interest only to western Europe. According to the session's rapporteurs, his comments provoked laughter amongst his colleagues.[44]

Again with regard to the 400th anniversary of the birth of the famous Pisan, it is instructive to read what Mgr Elchinger, coadjutor to the bishop of Strasbourg who then served as a kind of liaison with the academic community in France, reported in his memoirs. He recalled how he intervened, in November 1964, 'in the debate on the Church and culture to ask for the rehabilitation of Galileo'. His attempt at reconciliation 'was, at the beginning, very poorly perceived by some prelates but very well received by the pope Paul VI'. According to Elchinger, 'Galileo's trial had taken on the appearance of a "sin against the mind" and created a gap between the universe of faith and the universe of science'.[45]

During the Second Vatican Council, the French newspaper *Le monde* reported Elchinger's critical comments on the manner in which the Church continued to view the sciences. After being asked if the Church did not have 'a morbid fear of rationalism and the critical spirit', the Bishop replied that 'Galileo remains a symbol in the history of modern times. Let it not be said too quickly that it is part of ancient history. The conviction of Galileo has never been cancelled. Many scientists today attribute to the Church the attitudes of the theologians who four centuries ago condemned this great and honest scholar'. He recalled that 1964 'marks the fourth centenary of the birth of Galileo' and that it would be 'an eloquent act if the Church humbly accepted to rehabilitate him. The world of today expects of the Church something other than good intentions. It expects action'.[46]

A few months later, in firm support of his point of view, Father Dominique Dubarle, a French Dominican and philosopher of science, sent Archbishop Elchinger the text of a petition launched by the Union of French Catholic Scientists. With the support of hundreds of scientists, it recalled that there was still 'no status for the scholar guaranteeing him the freedom of research, regardless of his specialty'. The Catholic scholars deplored the fact that:

> Many intellectuals accuse the Church of a form of anachronistic dog-matism of which the condemnation of Galileo remains the symbol, all the more so that this condemnation has never been cancelled. They ask the Church if the celebration in this year of the fourth centenary of the birth of the great scholar would not be the opportunity to make such a gesture.[47]

Dubarle strongly supported the idea of a 'solemn rehabilitation of Galileo'. In March 1965, just before the Second Vatican Council agreed on the final wording of the text of the Pastoral Constitution, he explained himself at length to Elchinger, somehow charged with getting this message across to the cardinals. Dubarle was convinced that a 'sin against the mind' had been committed against Galileo, that the Church was wrong at the doctrinal level and that it had legally abused the Decree of 1616 during the trial of 1633. According to Dubarle, 'all this is public knowledge and continues to create scandal, especially in the scientific community'. But he warned that mere 'fine words, loudly proclaimed, on the esteem in which the Church holds science and scientists, or even the diplomatic regrets about the unfortunate incidents' would finally do 'more harm than good'.[48] In these circumstances, he saw only two possible approaches:

> It is necessary to either have the courage to leave the question in its present state, the decisions to be taken in this regard being as yet insufficiently mature from a Catholic point of view, or do now what will be sooner or later have to be done. By the 'necessary' I mean: (a) The solemn annulment of Galileo's trial of 1632–33 and the complete reformation of the judgment condemning him to abjure, (b) the express disavowal of certain procedures by the Holy Office and the public institution of forms of procedures guaranteeing the 'rights of human persons' brought before judges in eventual doctrinal trials within the Catholic Church.[49]

Of course, all this was too straightforward and too radical for the Church, still caught up in the internal strife that had led to compromise and the slowest change possible. The Cardinals were therefore divided on the exact way to recognize the historical wrongs of the Church, and the conservative faction prevailed once again, with the support of Paul VI, who preferred not to disturb the Galileo case. In effect, to frankly admit to having made an error would clearly affect the legitimacy of the Church as an institution. The debates and secret councils led eventually to the final formulation of paragraph 36 of the Pastoral Constitution on the Church in the Modern World, *Gaudium et Spes*. This text reflected once again the contortions the scribes of the Vatican bureaucracy were capable of to avoid saying something clearly all the while subtly suggesting the same so as to be able to claim later that what was asked for had already been granted.[50] After commenting at length on the legitimacy of a just autonomy for the sciences (to which we will return in chapter 5), the paragraph added that:

> we cannot but deplore certain habits of mind, which are sometimes also found among Christians, which do not sufficiently attend to the rightful independence of science and which, from the arguments and controversies they spark, lead many minds to conclude that faith and science are mutually opposed [62].

The 'end note 62' then referred the reader (without comment) to Paschini's book published by the Vatican in 1964. Unaware of all the changes made to his master's manuscript, Maccarrone concluded that Paschini had finally won out having received the rare honour of appearing by name in an Apostolic Constitution, something hitherto reserved for popes and the Holy Scriptures. He considered this 'solution' to the 'Galileo problem' a better outcome than a revision of the trial as some continued to demand after the Second Vatican Council.[51] Several outside observers also believed the problem to be solved. For example, in his book on *The Galilean Revolution*, published in 1969, the philosopher Georges Gusdorf claimed that Vatican II 'has put an end to this lamentable history, by ensuring that the Florentine scholar received the atonement due to him'.[52] But as Father Dubarle had foreseen, this half-measure, decoded only by insiders, could not meet the demand, repeatedly expressed for centuries, for a 'real', that is to say an honest, complete and direct, rehabilitation of Galileo.

The intervention of John Paul II

It was this insistent request for a less coy declaration that explains why John Paul II decided, at the end of the 1970s, to try to finally remove this pebble in the shoe of so many popes, one that had made walking difficult for them especially when they were heading in the direction of scientists.

Here again, a commemoration provided the opportunity. As there was no symbolic date in the offing to commemorate Galileo or Copernicus, it was the 100th anniversary of the birth of Albert Einstein that served as a pretext for Pope John Paul II to speak of Galileo before the Assembly of the members of the Pontifical Academy of Sciences on 10 November 1979. The pope clearly wished to go beyond the euphemisms of Vatican II. After having recalled that the Vatican II Council had 'recognized and deplored certain unwarranted interventions' and quoted the paragraph 36 mentioned above, he affirmed that the 'reference to Galileo is clearly expressed in the note to this text, which cites the volume *Vita e opere di Galileo Galilei* by Mgr. Pio Paschini, published by the Pontifical Academy of Sciences'. He then admitted that it was nonetheless necessary 'to go beyond this stand taken by the Council' and declared that Galileo's case should be analysed once again by 'theologians, scholars and historians, animated by a spirit of sincere collaboration' in order to finally 'dispel the mistrust that still opposes, in many minds, a fruitful concord between science and faith, between the Church and the world'. He said he would 'give all [his] support to this task, which will be able to honour the truth of faith and of science and open the door to future collaboration'.[53] As always, these few sentences hide the many conflicts within the Roman Curia that opposed those who believed that the case had been settled permanently by Vatican II and those who, on the contrary, demanded a more honest and direct acknowledgment of the errors of the Church. In this way, the mention of an apparently insignificant detail about a reference to Galileo in a footnote at the bottom of a page of the Pastoral Constitution *Gaudium et Spes* suggested that Galileo had indeed been mentioned even though the text was vague and said nothing specific, leaving only exegetes to decrypt the admission of an error with regard to the great scientist. Having recalled this detail to reassure those who believed the case settled, John Paul II asserted immediately after that it was necessary to go further and recognize 'wrongs from whatever side they come'.[54]

66

The Vatican bureaucracy having its own inertia, it took another eighteen months for the commission charged with reviewing the various facets (cultural, historical and theological) of Galileo's trial to be formally constituted. Its composition continued to reflect the search for compromise: presided by Bishop Garonne, who had been a supporter of Galileo during Vatican II, but now very old, the Commission included Michele Maccarrone, who represented to some extent his master Paschini, as well as the Jesuit Lamalle who, as we saw, was responsible for the 'revisions' made to Paschini's original work. Finally, the Cardinal Paul Poupard, President of the Pontifical Council for Culture, was also a member of the Commission; he would play a vital role in the final phase of its activities.

According to the Jesuit astronomer George V. Coyne, then Director of the Vatican Observatory and also a member of the Commission, the Commission did little, no longer meeting after the end of 1983 and produced little that was new to the already much discussed Galileo case. When the Pope asked the Chairperson of the Commission about the progress of their work in 1989, he learned, according to Coyne, that many of its original members were 'either deceased or no longer active'.[55] It was then decided that Bishop Poupard should prepare a final report for the Pope so that he might make a statement on the subject.[56] However, according to the most eminent Galileo specialists, the interpretation ultimately proposed in the report could not be seriously regarded as a synthesis of the Commission's work.

The irritation felt by numerous Galileo experts upon reading the document prepared by Poupard was due to the fact that the Cardinal appeared once again to avoid admitting clearly that the Church had committed an error in condemning both the doctrine of Copernicus and Galileo himself.[57] Poupard went so far as to praise Bellarmine, whom he presented as a better epistemologist than Galileo! He also repeated the fallacy that only 'irrefutable' proof would have justified Galileo's point of view and that such evidence was not gathered until much later, thus justifying Bellarmine's original stance. Far from appeasing, the conclusion sowed bitterness. Coyne himself wondered whether it was fair to talk, as the Pope did, of a 'Galileo myth' because 'it may be a genuine historical case of a continuing and real contrast between an intrinsic ecclesial structure of authority and the freedom to search for the truth in whatever human endeavor, in this case in the natural sciences'. Writing more than fifteen years after the Pope's famous 1992 speech that put an end to the investigation launched in 1979, Coyne still felt that Galileo should have had the freedom to continue his research, even assuming the falsity of the Copernican system. But this freedom was refused 'by official declarations of the Church'. And therein, Coyne concluded, 'lies the tragedy'.[58] However, in Poupard's work, these fundamental questions, as well as those raised by Dubarle thirty years earlier, were ignored.

Despite the reservations of the Galileo specialists – who were in fact never consulted about the conclusions to be drawn from the activities of the Commission[59] – John Paul II delivered a final speech on the subject to the Pontifical Academy of Sciences in October 1992, taking advantage of the opportunity offered by the 350th anniversary of the death of Galileo, a symbolic date that probably influenced the decision to put an end to the committee's reflections.[60] Whatever the limits of this re-examination of the Galileo file, which added nothing really new, John Paul II's gesture allowed the Roman Church to settle the question in a way that is probably definitive. Even Fantoli, a historian rather critical of the manner in which the work was conducted, admitted that one could not ignore that the Pope's discourse officially recognized the mistakes committed by the Church in 1616 and 1633.[61] Of course, these errors were presented in such a way as to suggest that both parties were at fault. Thus, according to the Pope, Galileo 'rejected the suggestion made to him to present the Copernican system as a hypothesis, inasmuch as it had not been confirmed by irrefutable proof. Such, however, was an exigency of the experimental method of which he was the inspired founder'. Historians have been quick to point out the fallacy here: the so-called 'scientific method' does not and did not, even in Galileo's time, require a proof that is 'irrefutable', but a rational comparison of the theses presented. For it is quite obvious that there was also no 'irrefutable' evidence of the *immobility* of the Earth! Conversely, Cardinal Bellarmine (Poupard's idol), 'had seen what was truly at stake in the debate' because he 'personally felt that, in the face of possible scientific proofs that the Earth orbited round the Sun, one should interpret with great circumspection every biblical passage which seems to affirm that the Earth is immobile'. For good measure, the Pope admitted that Galileo, 'a sincere believer, showed himself to be more perceptive' in matters of biblical interpretation 'than the theologians who opposed him'. He even admitted that Galileo's letter to the Duchess Christina of Lorraine, was 'like a short treatise on biblical hermeneutics'.

In sum, John Paul II concluded in his 1992 speech, the whole story was one of 'tragic reciprocal incomprehension', which 'has been interpreted as the reflection of a fundamental opposition between science and faith'. The clarifications made by recent historical studies, he added, 'enable us to state that this sad misunderstanding now belongs to the past'.[62] He thus sought to replace the long era of conflict with a new era of 'dialogue'.

Before examining (in chapter 5) the historical evolution of the discussion of the nature, confrontational or not, of the relations between science and religion since the beginning of the nineteenth century, we need first to trace out the long process that led to the complete secularization of the scientific method, a process that finished in the middle of the nineteenth century, by completely excluding from the domain of legitimate scientific explanation any recourse to miracles or other supernatural intervention to account for natural, historical or social phenomena. It is therefore to the slow implementation of the naturalist method in the different fields of science and to the correlative exclusion of God from the field of science that we now turn.

— 3 —

GOD: FROM THE CENTRE TO THE PERIPHERY OF SCIENCE

God, if we may express it in such a way, from being at first present in every human relationship, has progressively withdrawn. He leaves the world to men and their quarrels.

Emile Durkheim[1]

Modern science was institutionalized at the beginning of the seventeenth century by explicitly excluding any discussion of religion and politics from its organizations. Thus, the Charter of the Accademia dei Lincei, a scholarly society created in Rome in 1603 by Prince Federico Cesi (Galileo became a member in 1611), stipulated that members should not discuss politics or religion which could only engender passion and hatred among its members and that they should limit themselves to discussions related to the physical and mathematical sciences.[2] Similarly, half a century later, members of the Royal Society of London, founded in 1660, had to adhere to this principle of separation between science and politics and religion.[3] At the end of the century, taking advantage of an administrative reform, the regulations of the Paris Academy of Sciences went so far as to instruct its members that no one could be proposed to Her Majesty to fill a vacant position at the Academy if he were member of a religious order.[4] Two centuries later, the appearance of neutrality remained just as important and, once again to avoid conflict with religious authorities, the statutes of the Prehistoric Society of France, founded in 1904, stipulated in Article 2 that the Society forbids discussion of all matters beyond its purview especially 'all political or religious discussion'.[5]

The specific political, intellectual and social contexts of these times directly contributed to the establishment of boundaries defining legitimate speech within scholarly societies and delineating a

70

social space within which participants deployed a scientific discourse with its own norms.[6] Even the fear of censorship contributed to the exclusion from physics of any reference to what was seen as the prerogative of theologians. Thus, at the time when the theologians of the Sorbonne attacked the philosophy of Descartes, Jacques Rohault (1618–1672), the author of a 1671 Cartesian *Traité de physique*, which quickly became a reference book reprinted and translated into Latin and English, was aware of the danger of censorship.[7] Getting ahead of the game, he published in the same year his *Entretiens sur la philosophie* in which he reminded the reader that 'theology and philosophy have different principles; theology is based on authority and revelation and philosophy is based on reason alone, from which it follows that one can treat one without the other'.[8] As the Faculty of Theology 'discussed only matters related to faith', he insisted that he had taken 'all the necessary precautions' not to address these issues and had 'only dealt with things as they are in their ordinary and natural condition'. He even made 'an express declaration, repeated two or three times in different places, to the effect that he left it to the theologians to teach how things could be in an extraordinary or supernatural state'.[9]

We find the same concern for the separation of domains at the British Association for the Advancement of Science (BAAS), founded in 1831 for the promotion of increasingly specialized scientific disciplines (geology, physics, mathematics, etc.). Confronted with conflicts between the different Protestant denominations (Anglican, Quakers, Methodists, etc.), the BAAS also believed that it could achieve its objectives only by scrupulously avoiding religious and political debate. At the same time, the BAAS insisted on the fact that science and religion were not in conflict but contributed, each in their own way, to the glory of God.[10] The more conservative Anglicans, however, had a dim view of any attempt to separate science from religion. Immediately following the creation of the BAAS, John Henry Newman (1801–1890) – a future Catholic cardinal (he converted in 1845) – gave a sermon at Oxford denouncing the 'establishment of Societies in which literature or science has been the essential bond of union, to the exclusion of religious profession'. Although created with the best of intentions, these bodies have 'gradually led to an undue exaltation of the Reason and have formed unconstitutional power, advising and controlling the legitimate authorities of the soul'.[11]

As seen in the case of Rohault, and it would be easy to quote others, the *institutional* autonomy of the sciences had an *epistemological* counterpart, most often left implicit, that consisted in avoiding the constant invocation of God or miracles to *explain* natural phenomena. This epistemological separation between science and religion that began slowly at the beginning of the seventeenth century and that ensured scientific discourse a certain degree of autonomy – not without resistance from more conservative theologians – does not imply that scholars themselves no longer held religious beliefs. It was in fact a given that the overwhelming majority of scholars believed in God and his omnipotence and that many Jesuits, for example, were interested in and contributed to the advancement of knowledge of the natural world.[12] But science as conceived by Kepler, Galileo and Rohault, to name only a few, was neither theology nor metaphysics; it sought only to understand phenomena through constant and invariable laws of nature that were to be discovered by experimentation and mathematical demonstrations. In his reflections on the relationship between faith and reason, Newman clearly understood that men of science had often shown a certain tendency towards non-belief. As a result, he felt that:

> The system of physical causes is so much more tangible and satisfying than that of final [cause], that unless there be a pre-existent and independent interest in the inquirer's mind, leading him to dwell on the phenomena which betoken an Intelligent Creator, he will certainly follow out those which terminate in the hypothesis of a settled order of nature and self-sustained laws.[13]

With the return of an ecumenical discourse on the relationship between science and religion since the 1980s, innumerable articles have been written by historians of science that insist on the deep religious beliefs of the great scientists (Kepler, Newton, Faraday, Maxwell, Einstein, etc.),[14] as if this somehow proved that the idea of conflict between science and religion was only a myth forged by 'positivists' during the last quarter of the nineteenth century.[15] There is here, however, a serious methodological confusion, because most of these studies are biographical whereas the question of the conflict between science and religion is above all *institutional*. It is based on a conflict of authority between institutions with different aims and not on the psychology of individuals and the reasons that motivate them to undertake a scientific career and to reconcile – or not – their faith and their discoveries.

So as not to confuse scientists' personal religious convictions and the specific logic at work within a given scientific community endowed with its own specific rules of evidence and argument, it is useful to recall here the distinction proposed in the 1930s by the philosopher

72

of science Hans Reichenbach between what he called the 'context of discovery' – but would be more appropriately called the 'context of pursuit of research', as not all research leads to discovery – and the 'context of justification'.[16] While the context of pursuit of research provides all kinds of ideological, religious or psychological reasons to explain why a particular scientist pursues a given line of research or a specific hypothesis, the context of justification demands reasons deemed legitimate by the scientific community to accept a hypothesis, a theory or a research result. Fundamentally, Reichenbach's distinction points to two different levels. The first refers to individuals and the reasons for their choices, while the second refers to the institutional level of a community. For example, a researcher who has theological or more broadly religious reasons for studying nature falls within the context of research, whereas convincing one's colleagues that a cosmological model is valid falls within the context of justification and is based on theoretical and empirical arguments considered acceptable by the community of researchers at a given time. Thus, as we saw in the first chapter, Kepler, although a very pious Lutheran, did not hesitate to oppose theologians by reminding them that if, in theology, one had to take the views of authorities into account, only the weight of reason counted in philosophy.[17] Similarly, in the middle of the nineteenth century, a Christian evangelical such as the Scottish physicist James Clerk Maxwell, who had no doubt that God had created the laws of nature, was opposed to the public use of the results of science for religious purposes. For Maxwell, the harmony between science and religion was a personal matter.[18] Finally, even if Newton was convinced that his discovery of the law of universal gravitation confirmed the active presence of God in the universe, this did not prevent scholars in succeeding generations from sticking to his equations and leaving aside his religious interpretations. In sum, science *as an institution* effectively depersonalizes the content of science and makes it independent of the most personal interpretations, including those proposed by the discoverers themselves. Thus, Newton's absolute space was more useful to the physicists of the eighteenth and nineteenth centuries than his idea of the *sensorium Dei*. And even though this idea corresponded to Newton's profound religious convictions, natural philosophers and physicists did not have to adhere to it in order to use his *Principia*.

As we will see, the idea that science should propose only natural explanations within the framework of its own methodology, and exclude any appeal to miracles or divine intervention, was at work from the beginning of modern science. And it was so even though the expression 'scientific naturalism' that now serves to name this rule of method, only appeared in the middle of the nineteenth century in the context of the criticism that Christian evangelists raised against a science that had a tendency to forget about God.[19] This naturalism (more recently qualified as 'methodological') was first enforced gradually in the physical sciences and then in the geological and biological sciences, to the point where, in the middle of the nineteenth century, when these sciences had become specialized disciplines well established within the universities, it became a presupposition of science and simply went without saying. Alongside the sciences of nature, the study of history and civilization, which also emerged in the seventeenth century, cast a critical eye on religious texts, thus questioning the historical truth of the biblical story, ultimately replacing this miraculous story with a natural history of the men and the circumstances that produced this heterogeneous document.[20] Thus, after centuries of fierce resistance, most of the faithful of the Christian churches eventually came to recognize the autonomy of scientific discourse on the sky (cosmology), the Earth (geology), living beings (biology), and finally humans (anthropology and history).

In search of natural causes

In his *Dialogue Concerning the Two Chief World Systems*, one of the foundational books of modern science, Galileo suggested that it was best to keep to natural causes to explain physical phenomena. Against Galileo's thesis – represented in the book by the character Salviati – that the tides can be explained by the rotation of the Earth, Simplicio, his scholastic opponent, argued that he preferred to resort to miracles rather than accept this explanation that contradicted common sense. Salviati asked Simplicio why not simply consider that this movement of rotation could have itself been produced 'supernaturally' by a divine power? This would have the advantage, he added, not without irony, of avoiding multiplying miracles, for 'the miracle of making the water move brings another miracle in its train, which is that of holding the Earth steady against the impulse of the water. For these would be capable of making it vacillate first in one direction and then in the other, if it were not miraculously retained'.[21] Sagredo, an honest man open to new ideas, then intervened by suggesting that as to the miracle, one should have 'recourse to that only after we have heard arguments which are restricted within the bounds of nature'.[22] We clearly have here the idea that science explains phenomena through natural causes. God intervenes only in the selection of the 'initial conditions' of the universe. Salviati explained, for example, that one could 'suppose that among the decrees of the divine Architect' was the thought of creating the planets that we now observe and of assigning to each of them 'those degrees of velocity which originally seem good to the Divine mind'.[23] But once created, this world would obey immutable laws that would have to be discovered by the human mind.

Isaac Newton and Robert Boyle, two of the most important English natural philosophers of the middle of the seventeenth century are invoked more often than Galileo to demonstrate that science and religion can be harmonized and may even be mutually reinforcing. However, while it is true that Boyle attacked 'atheists' and defended the Christian religion in many publications, he did so in texts separate from his scientific writings that sought instead to describe empirical *'matters of fact'* and to avoid metaphysical speculation. His style of writing tended to accentuate the autonomy of science with regard to theology and Christian apologetics, disciplines regarded as having objects and functions different from those of natural philosophy.[24] Indeed, one seeks in vain in his many empirical studies for an invocation of God to *explain* his observations. His proposed explanations were based instead on the laws of nature that he sought to highlight. A promoter of the 'mechanical philosophy', Boyle believed that once created by God, the universe worked as a vast machine whose operation required explanation.[25] According to the logic of 'natural theology', once the observed phenomena had been described and explained, the reader could still believe that the complexity of the world or its uniformity attested to the greatness of God. But statements like this always appeared in the preface or conclusion and were not part of the *explanatory logic* proper to natural philosophy; the word God only entered the discourse when used in formulas like 'Thank God' or 'God willing'.

Between 1665 and 1700, less than three articles per year, on average, containing the word 'God' can be found in the *Philosophical Transactions* of the Royal Society of London, a number that decreased to less than two articles per year during the next century.[26] Henry Oldenburg, for example, the founder of the journal, the first of its kind, concluded his opening dedication to the members of the Royal Society, by wishing that 'the Great God prosper you in the noble engagement of dispersing the true lustre of his glorious works and the happy inventions of obliging men all over the world, to the general benefit of mankind'.[27]

Natural theology: science in the service of God

The idea that science and theology, far from being opposed, are compatible or complementary, lies at the root of natural theology. This tradition of thought, which invokes the order and the layout of nature to demonstrate the existence of God, can be traced back to Plato. The Greek philosopher claimed in *Laws* that there are two kinds of reasoning that lead to the belief in the gods, one of which is 'the ordering of the motion of the stars and all the other bodies under the control of reason'. And no man 'that views these objects in no careless or amateurish way has ever proved so godless as not to be affected by them'.[28]

Natural theology found particularly fertile ground in England at the end of the seventeenth century, where it was closely linked with the development of the sciences. The discipline's explicit function was to invoke scientific discoveries to show the action of divine power, the only possible source of the order and beauty of nature. The basic argument of natural theology, which remains essentially unchanged even today, was that the existence of the watch supposes and implies the existence of a watchmaker. Among the many reference works, constantly republished and translated, are those of the English naturalist John Ray, who published *The Wisdom of God Manifested in the Works of the Creation* in 1691, and William Derham's, *Physico-theology, Or, A Demonstration of the Being and Attributes of God from His Works of Creation*, a product of the *Boyle Lectures* of 1712. A century later, in 1802, the Reverend William Paley's *Natural Theology or Evidences of the Existence and Attributes of the Deity* became a bestseller that occupied the intellectual scene in the middle of the nineteenth century. The continent also underwent a vogue for natural theology that saw the publication of an *Insecto-theology: Or a Demonstration of the Being and Perfections of God, from a Consideration of the Structure and Economy of Insects* (1738) by the German naturalist and theologian Friedrich Christian Lesser (1692–1754). In France, the abbot Noël-Antoine Pluche (1688–1761) made natural theology known to a wide public at the beginning of the 1730s in the many volumes of his *Spectacle de la nature*, translated into English as *Nature display'd: being discourses on such particulars of natural history as were thought most proper to excite the curiosity and form the minds of youth*.[29]

A founding member of the Royal Society of London, Robert Boyle provided an important impetus to natural theology by willing a part of his fortune to the creation of the *Boyle Lectures*, a cycle of conferences given in two London Anglican churches, St. Paul and St. Mary-le-Bow. Often published, the lectures promoted the Christian religion against 'infidels' and 'atheists' and asserted the compatibility of religion and science.[30] The curious obsession of Boyle and his friends with 'atheists' should not be taken as an indication that such persons actually existed in large numbers. 'Atheist' was then an insult aimed at discrediting competing conceptions of Christianity. Indeed, no writers of the time would have described themselves as 'atheist'.[31]

Inaugurated in 1692 with a conference by a friend of Newton, the Reverend Richard Bentley, on 'A Confutation of Atheism', the yearly *Lectures* continued until the end of the eighteenth century.[32] Following a decline, natural theology resumed its importance in England during the 1830s with the publication of a series of eight books, known as *The Bridgewater Treatises On the Power, Wisdom and Goodness of God As Manifested in the Creation* that sought to demonstrate how recent scientific discoveries concurred with the Christian religion. This renewed interest coincided with the increased social visibility of science in England through the activities of the BAAS (founded in 1831), and declined once again at the end of the nineteenth only to re-emerge a century later, at the end of the twentieth century, as we will see in chapter 6.

Like Boyle, Newton was also very religious and wrote as much if not more on theology and Christianity than on natural philosophy and mathematics. He was furthermore convinced that his research on the system of the world confirmed the active presence of God in the universe. But we often forget that the first editions of his two major scientific works, *Philosophiae Naturalis Principia Mathematica*, published in Latin in 1687, and the *Opticks*, published in English in 1704, never invoked God as the direct cause of the many phenomena that he sought to explain through natural laws. At the time, it went without saying that the universe and its laws were a divine creation, but it was implicitly recognized by most scholars that science was committed to the discovery and description of secondary causes. Newton himself, who always kept his deep Unitarian religious beliefs – 'non-conformist' from the point of view of Anglicans – to himself and who never in his lifetime published his theological research, stated that 'religion and Philosophy are to be preserved distinct. We are not to introduce divine revelations into Philosophy nor philosophical opinions into religion'.[33] As Peter Bowler noted, 'Newton's practices involved both his science and his religion, but he could reveal only the scientific aspects of his researches without risk to his career'.[34] This is a way to recall the difference between the individual convictions of scientists and the collective nature of science as a social institution.

Newton did not, however, always follow his principle and added, in the second edition of his *Principia* (1713), published twenty-five years after the first, a *General Scholium* at the end of the book as a general conclusion. We find here his famous invocation of a divine action to ensure the stability of the solar system, which, he believed, would in principle disperse if submitted to the sole action of the gravitational force between the planets. He thus clearly introduced into his work arguments drawn from the 'argument from design' central to natural theology. This famous argument remains to this day central as 'evidence' for the existence of God. The few paragraphs that relate directly to God and His nature end abruptly with the assertion that God is actually part of natural and experimental philosophy: 'this concludes the discussion of God, and to treat of God from phenomena is certainly a part of experimental philosophy', the latter term being changed to 'natural philosophy' in the third, 1726 edition.[35] After this sentence, Newton returned immediately to the question of the nature of gravitation, the force that explained celestial phenomena, and admitted to being unable to assign it a mechanical cause. As Stephen Snobelen has noted, 'the fact that Newton finds it necessary to assure his readers in the General Scholium that discoursing of God does belong to the domain of experimental or natural philosophy does nevertheless demonstrate both that the assumption was already in dispute and that Newton and others were perfectly able to articulate a distinction between these fields of endeavour'.[36]

Until this second edition of 1713, Newton had admitted in practice that natural philosophy aimed above all to explain phenomena by natural causes, having been convinced that this approach served the interests of religion. The late addition of the *General Scholium* might be explained as a response to the criticisms of scientists like Leibniz, who thought that the *Principia* could be interpreted as support for atomistic pantheists for whom nature could be explained without any divine intervention.[37] Although this was clearly not Newton's point of view – in his private correspondence he endorsed the arguments of natural theology – it was also true that in setting out the mathematical laws from which he deduced the movement of all the planets, he might have given the impression that God no longer actively intervened in nature.[38] Many of Newton's contemporaries thought that experimental philosophy was dangerous for religion, the State and society in general.[39]

The use of divine action as an explanation of what Newton considered a problematic instability of the solar system resembles the 'God of the gaps' argument invoked in the twentieth century by creationists to explain facts that they consider unexplainable by the usual scientific methods. Simplicio did the same in Galileo's *Dialogue* to account for the phenomenon of the tides while rejecting, at the same time, the movement of the Earth. At the epistemological level, these explanations are *ad hoc* and therefore arbitrary and came to be increasingly rejected by scientists in different disciplines – astronomers, physicists, naturalists and geologists – who were influenced, in the second half of the eighteenth century, by the rationalist current of the Enlightenment. Such scientists pushed the notion of divine intervention in nature to the periphery of science, no longer considering it a legitimate scientific explanation. By the middle of the seventeenth century, Spinoza had already undertaken his critique of the finalism at the basis of natural theology by noting that 'the will of God [is] the refuge for ignorance'. He had understood that this 'new mode of argumentation', which replaced 'demonstration by the absurd' with 'demonstration by ignorance', clearly served the interests of its promoters for they knew 'that if ignorance be removed, amazed stupidity, the sole ground on which they rely in arguing or in defending their authority, is taken away also'. Theologians and other promoters of natural theology thus declared 'heretical and impious' any person 'who endeavours to find out the true causes of miracles and who desires as a wise man to understand nature and not to gape at it like a fool'.[40]

As we saw, these criticisms did not prevent the spread of teleological arguments in the eighteenth century. Buffon, irritated to hear of the preaching of a theology of insects, criticized finalist explanations noting that 'the creator was great enough through His works' and was not made greater 'by our imbecility'. As he wrote in his *Discourse on the nature of animals*, those who consider that God created the universe and founded nature 'on invariable and perpetual laws' have a larger view of the 'Supreme being' than those who 'want to find him conducting a republic of flies and worried about the way to bend the wing of a beetle'.[41]

The writings of natural theology proclaiming that scientific discoveries could only confirm the greatness of God entertained a certain degree of ambiguity. In the 1830s, John Henry Newman questioned their usefulness and noted that they rendered a form of tribute to the Creator that had little practical utility for the believer. Those who already had faith could indeed find in natural phenomena a reason to admire the great wisdom of God, but Newman considered that this natural theology would hardly prevent individuals from abandoning religion or help in converting unbelievers.[42] He even wondered if atheism was not a philosophy that was just as compatible with the phenomena of the physical world as the doctrine of a divine active and creative power.[43] Converted to Catholicism, Newman continued to mistrust natural theology. In his sermons as Rector of the Catholic University of Dublin in the middle of the 1850s, he explained that 'the God of Physical Theology may very easily become a mere idol' and might even dispose the scientist against Christianity 'because it speaks only of laws and cannot contemplate their suspension, that is, miracles, which are of the essence of the idea of a Revelation'. That God was thus 'not very different from the God of the Pantheist'. He even wondered if he did not 'prefer that such a scientist adhering to natural theology should be an Atheist at once than such a naturalistic, pantheistic religionist' for his 'profession of Theology deceives others, perhaps deceives himself'.[44]

God: a useless hypothesis

The epistemological position defended by the French scientist Pierre-Simon Laplace (1749–1827), considered the 'Newton' of the end of the eighteenth century, is an important example of the explicit exclusion of God from the field of legitimate explanations of natural phenomena. There is a story, considered plausible, according to which Laplace responded to Napoleon, who wondered where God was in his book, by saying that he had no need of that hypothesis.[45] In his *System of the World*, originally published in 1796 and reprinted many times, Laplace even felt obliged to correct his illustrious precursor on this point. Having emphasized the importance of seeking only in the 'primordial laws of nature the cause of phenomena' he could not help but remark:

> how Newton has erred on this point, from the method which he has otherwise so happily applied. Subsequently to the publication of his discoveries on the system of the world and on light, this great philosopher abandoned himself to speculations of another kind, and inquired what motives induced the author of nature to give to the solar system its present observed constitution.

Newton had argued that 'this admirable arrangement of the Sun, of the planets, and of the comets, can only be the work of an intelligent and most powerful being.' However, Laplace wondered:

> could not this arrangement of the planets be itself an effect of the laws of motion; and could not the supreme intelligence which Newton makes to interfere, make it to depend on a more general phenomenon such

as, according to us, a nebulous matter distributed in various masses throughout the immensity of the heavens?[46]

A century after Newton, physics and astronomy had abandoned these external causes that were, Laplace concluded, 'only an expression of our ignorance of the true causes.'[47]

Geology versus The Deluge

The naturalist trend in science manifested itself not only in the physical and astronomical sciences. It also surfaced in other areas likely to come into conflict with a literal reading of the Bible, the dominant reading within the Christian world until at least the middle of the nineteenth century. The many botanical, zoological and ethnographic discoveries produced by geographical explorations from the sixteenth to the nineteenth century upset the vision of a fully known, static world. If Lactantius, whom Copernicus and many others after him cited as an example of the ignorance of science,[48] was still able, at the end of the third century, to deny the existence of peoples living in the antipodes under the pretext that they would have to walk on their heads, such a belief had become impossible by the sixteenth century. The application of a rational mode of thought inevitably undermined the more naïve religious conceptions of nature. It was indeed *a priori* unlikely that unexpected discoveries encountered so far, and those to come, would easily be made compatible with biblical texts. At the same time as the Catholic Church quietly removed the books of Copernican astronomy from the List of Prohibited Books in the middle of the 1830s, geology had made great progress that called into question the existence of the flood described in the Bible.[49]

While the British naturalist John Woodward (1665–1728) could still at the end of the seventeenth century, in his book on the natural history of the Earth, invoke the direct action of God to explain many natural phenomena, this approach was already under attack at the beginning of the next century. In a book on marine fossils that appeared in 1721, the Italian naturalist Antonio Vallisneri remarked 'how much the interests of religion, as well as those of sound philosophy, had suffered by perpetually mixing up the sacred writings with questions in physical science'.[50] Twenty-five years later, his compatriot Cirillo Generelli suggested that it was not 'reasonable to call the Deity capriciously upon the stage, and to make him work miracles for the sake of confirming our preconceived hypotheses'. The naturalist was

81

rather trying to explain 'how these marine animals were transported into the mountains by natural causes'.[51]

In his *Natural History*, whose thirty-six volumes started appearing in 1749, the French naturalist Buffon, in proposing a general theory of the formation of planets and the evolution of the Earth, made no mention of the action of God except as a first cause. This prime mover was then quickly relegated to the periphery, leaving nature to act according to its own immanent laws. Describing the formation of the solar system with the help of Newton's theory of gravitation, Buffon noted that the unknown original impulsive force 'was unquestionably at first communicated to the planets by the Supreme Being', but immediately added that:

> in physical subjects, we ought, as much as possible, to avoid having recourse to supernatural causes; and, I imagine, a probable reason may be assigned for the impulsive force of the planets, which will be agreeable to the laws of mechanics, and not more surprising than many revolutions that must have happened in the universe.[52]

And while Buffon admitted that John Woodward 'has the merit of collecting many important observations', he did not hesitate to emphasize that his system 'added to the miracle of the universal deluge many other miracles, or, at least, physical impossibilities, which accord neither with the Scriptures, nor with the principles of mathematics and of natural philosophy', referring here to the title of Newton's book published eight years before Woodward's.[53] Buffon also criticized the naturalist Johann Jakob Scheuchzer (1672–1733), who had proposed in 1708 a diluvial theory of the Earth's history consistent with the biblical story, for mixing physics with theology.[54] Buffon asked:

> Why then would we suppose, with many naturalists, that the waters of the deluge totally changed the surface of the globe, even to the depth of two thousand feet? Why imagine that the deluge transported those shells, which are found at the depth of seven or eight hundred feet, immersed in rocks and in marble? Why refer to this event the formation of hills and mountains? And how is it possible to imagine, that the waters of the deluge transported banks of shells of 100 leagues in length? I perceive not how they can persist in this opinion, unless they admit a double miracle, one to create water, and another to transport shells. But as the first only is supported by holy writ, I see no reason for making the second an article of faith.[55]

Buffon's finesse was not to deny the existence of the Deluge but simply leave it in the domain of religion and faith. He considered that 'the deluge ought to be regarded as a supernatural mode of chastising the wickedness of men, not as an effect proceeding from any natural cause'.[56] It thus destroyed only humanity 'but it produced no change on the surface of the Earth', which 'before the deluge was nearly the same as now'.[57]

More deist or sceptic than a true atheist, Buffon nonetheless denounced the confusion of genres among those who:

> have embraced this hypothesis [of the universal deluge] with so blind a veneration, that their only anxiety is to reconcile it with holy writ; and, in place of deriving any light from observation and experience, they wrap themselves up in the dark clouds of physical theology, the obscurity and littleness of which derogate from the simplicity and dignity of religion, and present to the sceptic nothing but a ridiculous medley of human conceits and divine truths.

Buffon insisted that one 'must, above all, avoid blending bad philosophy with the purity of divine truth'.[58]

Despite Buffon's efforts to save both the biblical narrative and the autonomy of his science by not confusing genres, his purely natural explanations of the formation of the Earth and of the probable fate of the Sun were clearly incompatible with the biblical narrative. We will see in the next chapter that Buffon was able to avoid the Index of Prohibited Books only by maintaining that he considered his history to be merely a hypothesis, in accordance with the 'Catholic epistemology' of Cardinal Bellarmine.

After Buffon, geology developed on a strictly naturalist basis, avoiding any discussion of divine intervention as a legitimate scientific explanation. Charles Lyell's (1797–1875) *Principles of Geology*, published in three volumes at the beginning of the 1830s and constantly republished and updated throughout the century, quickly became a reference work for all scientists in the field. Geology then became increasingly specialized, and one searches in vain in Lyell for references, other than historical, to miracles or to other divine explanations of terrestrial phenomena. In addition, following the tradition inaugurated by Buffon and in the wake of James Hutton (1726–1797), Lyell's theory of the Earth's history was based on the idea that the forces acting on the Earth and in its interior were the same today as in the past. This approach opened the door to an Earth of great antiquity, an idea already suggested by Buffon, which was incompatible with that of a unique creation going back less than 10,000 years. By invoking forces that act in nature according to laws that were immutable in time and space, Lyell implicitly excluded any miracle or supernatural intervention as a possible explanation of geological phenomena. Catastrophist interpretations, on the other hand, continued to allow for divine intervention. During exceptional events like the Deluge, specific and discontinuous forces, rather than universal and invariable laws, intervened.

Reviewing the history of the geological theories invoking the Deluge as an explanatory factor, Lyell noted that 'never did a theoretical fallacy, in any branch of science, interfere more seriously with accurate observation and the systematic classification of facts'.[59] Fortunately, he added, things had changed and geology was 'no longer required to propel the vessel against the force of an adverse current.'[60]

One can easily understand that, once established, this mode of rational reasoning based on observation and the analysis of natural forces would come to embrace all natural phenomena. For natural history investigated not only the geographical distribution of minerals, plants and animals, but also human beings themselves, thus joining anthropology and prehistory. It was only a matter of time until the dynamics of naturalistic scientific research raised questions about the links between the species. This inevitably collided with the biblical belief that all species had been created directly by God.

The Darwinian Revolution

When the young Charles Darwin (1809–1882) set out on his voyage around the world in 1832, he took with him the first volume of Lyell's *Principles of Geology*, which helped him to understand the diversity of the species that he met in the course of his long journey. His fundamental idea of natural selection – which was sufficiently 'natural' for at least one other naturalist, Alfred Russel Wallace, to discover it independently in his own research in the Amazon and in the Malay Archipelago – constituted a crucial step in the application of methodological naturalism to natural history. Darwin explicitly recognized the intimate link between Lyell's geology, which implied a long history of the Earth, and the idea of the evolution of species. In his *On the Origin of Species*, which became famous upon its publication in 1859 and was immediately denounced as godless by Christians still attached to the literal reading of the Bible, Darwin said:

> I am well aware that this doctrine of natural selection [. . .] is open to the same objections which were at first urged against Sir Charles Lyell's noble views on 'the modern changes of the Earth, as illustrative of geology'; [. . .] and as modern geology has almost banished such

views as the excavation of a great valley by a single diluvial wave, so will natural selection, if it be a true principle, banish the belief of the continued creation of new organic beings, or of any great and sudden modification in their structure.[61]

Darwin also asserted that 'He who can read Sir Charles Lyell's grand work on the Principles of Geology, which the future historian will recognise as having produced a revolution in natural science, yet does not admit how incomprehensibly vast have been the past periods of time, may at once close this volume'.[62]

Acutely aware that this naturalistic approach applied to living beings could still not be taken for granted and that his book might raise theological objections, Darwin sought Lyell's advice, only a few months before publication, on how to assure his editor that his book was not 'more unorthodox than the subject makes inevitable'. Should he tell his editor that he did 'not discuss origin of man' nor 'bring in any discussions about Genesis', or was it better to say nothing and 'assume that he cannot object to this much unorthodoxy, which in fact is not more than any Geological Treatise, which runs slap counter to Genesis'?[63] Lyell quickly assured him that he had nothing to fear.

In any event, as a precaution, Darwin quoted authors known for their religious orthodoxy in the epigraph to the *Origin* such as the mathematician, philosopher and Anglican pastor William Whewell. The author, in 1833, of one of the eight Bridgewater treatises devoted to natural theology, Whewell had argued that 'with regard to the material world, we can at least go so far as this – we can perceive that events are brought about not by insulated interpositions of Divine power, exerted in each particular case, but by the establishment of general laws'. For Whewell, this 'is the view of the universe proper to science, whose office it is to search out these laws'. He adds that 'God is the author and governor of the universe through the laws which he has given to its parts, the properties which he has impressed upon its constituent elements'.[64] Although Darwin did not cite this part of Whewell's treatise which mentioned God explicitly, the appeal to the authority of the Reverend was designed to confirm the legitimacy of Darwin's view of science that in no way called into question Christianity, but simply affirmed the specificity of a science that sought to explain natural phenomena through natural causes. To those who might object that science had boundaries beyond which it should not venture and that the explanation of the origin of the species did not fall within its purview, Darwin was ready with a quote from the philosopher Francis Bacon, who still symbolized for many scientists the pinnacle of the experimental method. 'Let no man' Darwin cited from Bacon, 'out of a weak conceit of sobriety, or an ill-applied moderation, think or maintain, that a man can search too far or be too well studied in the book of God's word, or in the book of God's works; divinity or philosophy; but rather let men endeavour an endless progress or proficience in both.'[65] The reference to the two major books – those of religion and nature – took up an old analogy used by Galileo. The two 'books' may have the same source but, it was implicitly suggested, the method used to read the second is not the same as for the first. As will be seen in chapter 5, this comparison of the two books from the same 'author' will often be taken to prove that there can be no real conflict between science and religion.[66]

At the end of his book, Darwin reiterated his conviction that 'it accords better with what we know of the laws impressed on matter by the Creator, that the production and extinction of the past and present inhabitants of the world should have been due to secondary causes, like those determining the birth and death of the individual'.[67] All forms of life have been 'produced by laws acting around us'. As to the origin of life itself, it remained unknown and Darwin was convinced that life had been 'originally breathed into a few forms or into one'.[68]

In response to the numerous critics who considered his book irreligious and offensive, Darwin changed the indirect style of this sentence (and others) – which suggested a force immanent to nature rather than an active god and creator – by adding 'by the Creator' after 'originally breathed' in the second edition, published a few months later. He also added, at the beginning of the conclusion, a sentence stating that he saw 'no good reason why the views given in this volume should shock the religious feelings of any one'.[69] In the third edition of 1861, he also added another sentence in order to show 'how transient such impressions [of religious offense] are'. He recalled that 'the greatest discovery ever made by man, namely, the law of the attraction of gravity, was also attacked by Leibniz, "as subversive of natural and inferentially of revealed religion"'. And he confided that 'a celebrated author and divine' (it was Charles Kingsley[70]) had written to him that he had 'gradually learnt to see that it is just as noble a conception of the Deity to believe that He created a few original forms capable of self-development into other and needful forms, as to believe that He required a fresh act of creation to supply the voids caused by the action of His laws'.[71]

Toward the end of his life, in his *Autobiography*, Darwin stated that 'the more we know of the fixed laws of nature the more incredible do miracles become'.[72] Though the sentence was censured by his family in the first published version of his *Autobiography*, he clearly stated that 'everything in nature is the result of fixed laws'.[73] As for 'the old argument from design in Nature, as given by Paley, which formerly seemed to [him] so conclusive', it fails, 'now that the law of natural selection has been discovered'.[74] He had already, in *The Origin of Species*, criticized explanations based on final causes and noted that 'It is so easy to hide our ignorance under such expressions as the "plan of creation," "unity of design," &c., and to think that we give an explanation when we only restate a fact'.[75]

A natural history of Man

Although Darwin's revolutionary work was limited to the study of the transformations undergone by animal species and did not touch upon the question of human beings, considered at the time as having been directly created by God, his readers were perceptive enough to understand that there was no reason that natural selection or some other evolutionary force also acted on human beings. Darwin himself recalled in his Autobiography that as soon as he had 'become, in 1837 or 1838, convinced that species were mutable productions, [he] could not avoid the belief that man must come under the same law'.[76]

The first great theorist of a natural explanation of the transformation of species, Jean-Baptiste Lamarck (1744–1829), had also raised this question, even though his discussion was purely hypothetical and written in the conditional tense. The first part of his *Philosophie zoologique* of 1809, relating to the natural history of animals, ended with a section devoted to 'Some observations regarding to Man'. Lamarck noted that 'if man was only distinguished from the animals by his organisation, it could easily be shown that his special characters are all due to longstanding changes in his activities and in the habits which he has adopted and which have become peculiar to the individuals of his species'.[77] Consequently, it was easy to imagine that 'if some race of quadrumanous animals, especially one of the most perfect of them were to lose, by force of circumstances or some other cause, the habit of climbing trees and grasping the branches with its feet in the same way as with its hands' and were thus 'forced for a series of generations to use their feet only for walking, and to give up using their hands like feet', there was no doubt in Lamarck's mind that 'these quadrumanous animals would at length be transformed into bimanous, and that the thumbs of their feet would cease to be separated from the other digits, when they only use their feet for walking'.[78] Having 'acquired an absolute supremacy over all the rest' this 'predominant race', would thus 'ultimately establish a difference between itself and the most perfect animals', and indeed 'leave them far behind'.[79] To emphasize the purely hypothetical character of his remarks, Lamarck closed the section with the statement: 'Such are the reflections which might be aroused, if man were distinguished from animals only by its organisation, and if his origin were not different from theirs'.[80]

Clothed in a hypothetical form of presentation, these conjectures do not seem to have generated any official reaction on behalf of the Church or on behalf of more conservative Catholics.[81] In addition to the author's caution, it should be added that revolutionary France of the beginning of the nineteenth century was quite rationalist and little inclined to denounce scientists as in Buffon's time. But this did not prevent several contemporary observers such as the naturalist Julien-Joseph Virey from asserting that Lamarck's views led directly to atheism.[82] Be that as it may, Lamarck's 'observations' showed that the naturalist mode of thought could easily be applied to the origins of man. And Darwin did recall in the third edition of his *On the Origin of Species* that Lamarck was the first to render to science 'the eminent service of arousing attention to the probability of all change in the organic as well as in the inorganic world being the result of law, and not of miraculous interposition'.[83]

The debate on the origins of humans finally broke out in the middle of the nineteenth century as a result of numerous archaeological discoveries that clearly highlighted their antiquity, and thus, once again, called into question the lessons of Genesis that had been taught for 2,000 years in the Christian world. By the time *On the Origin of Species* had appeared, archaeological research had already unearthed carved stones strongly suggesting the antiquity of the human species. In France, Jacques Boucher de Perthes (1788–1868) had begun accumulating 'antediluvian axes' at the beginning of the 1840s in Abbeville, Picardy, and had made his discoveries known in 1847 in his *Antiquités celtiques et diluviennes*. First sceptical as to the value of the discoveries by this amateur, the most well-known geologists and archaeologists in Europe, including Charles Lyell in 1859, visited the sites and confirmed their authenticity.[84] From flint axes, it was a simple step to search for the people who made those axes and Boucher de Perthes went on to publish work dealing directly with humans, including *De l'homme antédiluvien et de ses œuvres* (1860) and *Nouvelles recherches sur l'existence de l'homme et des grands mammifères fossiles* (1862).

Lyell added his authority to these ideas in 1863 by publishing *Geological Evidences of the Antiquity of Man*, a synthesis of the discoveries reported in many European countries (Germany, England, Belgium, Denmark, France and Switzerland). Conscious of the fact that his strictly naturalist analysis of the origins of Man might be viewed as a form of a materialist philosophy, Lyell, who became a supporter of the theory of evolution, concluded his book by answering the charge of 'materialism brought against all forms of the development theory'. For him, 'far from having a materialistic tendency, the supposed introduction into the Earth at successive geological periods of life – sensation, instinct, the intelligence of the higher mammalia bordering on reason – and lastly the improvable reason of Man himself, presents us with a picture of the ever-increasing dominion of mind over matter'.[85] Lyell thus declared himself more spiritualist than materialist, since human intelligence, as well as the ideas that it engendered, was the fruit of a long natural evolution.

A natural history of religion

The naturalist mode of thinking at work in nineteenth-century geology and the other sciences that studied the history of the Earth and that attempted to explain observed changes clearly affected the study of religion and the historical sources of the Bible. The development of critical history and philology raised questions about how the biblical texts had been brought together and about the historical or mythical character of the assertions they contained, as well as their evident inconsistency. This is not the place to review this long history, equally as rich in conflict and religious condemnations.[86] But to complete this overview of the progressive exclusion of God from scientific reflection on natural and social phenomena, we need to briefly recall that historians also adopted an increasingly naturalistic approach to the past and ceased, in the course of the nineteenth century, to use the notions of miracle and providential intervention as explanatory factors of historical events. In 1835, for example, the German historian David Friedrich Strauss (1808–1874) offered an interpretation of the history of Jesus that allowed him to account for his 'miracles' without involving supernatural phenomena. His book, *The Life of Jesus Critically Examined*, considered materialistic if not frankly atheist, created a scandal and after its release he was forced to abandon his position as a teacher in a Protestant seminary in Tübingen.[87] Nonetheless, the book was a resounding success and went through several editions and translations. Even before its English translation, Philip Harwood, a London Unitarian pastor, critical of miracles, published a series of lectures on *German Anti-Supernaturalism*, presenting Strauss's analysis.[88]

The French philologist and historian of religion, Ernest Renan (1823–1892), was also an active spokesperson for the new critical history that had first emerged in Germany. Like Strauss, he wrote a *Life of Jesus* (1863) that was very popular and that also cost him his position at the Collège de France.[89] In an article entitled 'The critical historians of Christ' first published in 1849, he clearly summarized the method of critical history as applied to the miraculous phenomena recounted in religious stories:

> Criticism has two modes of attacking a marvellous narrative; for as to accepting it as it stands, it cannot think of it, since its essence is denial of the supernatural. 1. It may admit the substance of the narrative, but explain it by taking into consideration the age, the persons who have transmitted it to us, and the forms adopted at such and such an epoch to express facts. 2. It may cast doubt on the narrative itself, and may account for its formation without conceding its historical value. On the first supposition the task is to explain the material of the history; the reality of the material is consequently assumed. On the second, without pronouncing on this reality, the phenomena of the narrative are analysed as simple psychological facts. It is regarded as a poem created in all its parts by tradition, and neither having nor requiring other cause than the instincts of man's spiritual nature.[90]

A few years later, in 1857, in the preface to a collection of his papers on religious history, Renan returned to the opposition between historians and theologians and pointed out that 'a grave difficulty attaches to these studies, and timid people are led to ascribe to the writers who take them up, tendencies and aims to which they are strangers. It is in the nature of religions to exact an absolute credence, consequently to set themselves above common rights, and to deny the impartial historian's competency to judge them'. And 'in order to support their claim to be exempt from all reproach, [religions] are obliged to have a special philosophy of history, founded on a belief in the miraculous intervention of Deity in human affairs – an intervention whose benefits accrue to them alone'.[91] He insisted, moreover, that the first principle of historical criticism was 'that miracle has no more place in the tissue of human affairs, than it has in the series of natural facts' because 'every event in history has its human explanation, even though the explanation escape us through lack of sufficient knowledge'. It followed that there could be no common ground for historians and theologians, because the latter 'employ a method opposed to its own, and pursue a different end.' Renan reminded his readers that he never broached 'the fundamental question on which religious discussion turns, that is to say, the question of fact in regard to revelation and the supernatural', for this was not his concern.[92] He added that 'sensitive, like all powers that claim for themselves a divine source, religions naturally construe as hostility, even the most respectful expression of difference, and see enemies in all who exercise on them the simplest right of reason'.[93] He concluded that it was better to 'allow religions to proclaim themselves unassailable, since otherwise they will not gain due respect from their adherents; but let us not subject science to the censorship of a power that has no scientific character'.[94] As we will see in the next chapter, this book and several others from his pen dealing with the history of religion will be placed on the Index, as was also the case for Strauss's *Life of Jesus*.

Naturalizing God

Darwin was not a historian of religion like Strauss or Renan, but by retracing the natural history of man he would ultimately address these issues with a naturalist's eye. Having initially kept his reflections on the origins of man to himself, Darwin made them public in 1871 in *The Descent of Man*. From the outset, he recalled that although he had accumulated notes on the descent of man for many years, he had decided not to publish anything, convinced that doing so would 'only add to the prejudices against [his] views' on natural selection. But twenty years after the publication of *On the Origin of Species*, he believed that the context had changed and that his ideas had been fairly well accepted by naturalists. This new situation had encouraged him 'to put together [his] notes, so as to see how far the general conclusions arrived at in [his] former works were applicable to man'. He noted that 'The high antiquity of man has recently been demonstrated by the labors of a host of eminent men, beginning with M. Boucher de Perthes; and this is the indispensable basis for understanding his origin'. He, therefore, took 'this conclusion for granted'.[95] Drawing a lesson of epistemological optimism from the many recent archaeological discoveries concerning the human species, he remarked in passing that 'it has often and confidently been asserted, that man's origin can never be known: but ignorance more frequently begets confidence than does knowledge; it is those who know little, and not those who know much, who so positively assert that this or that problem will never be solved by science'.[96]

In studying the origin of humanity from a purely naturalist point of view, not only did Darwin never invoke 'God' as an explanatory cause, but he went even further by suggesting that the belief in God was itself the result of evolution by natural selection. The first chapter of *The Descent of Man* brought together the 'evidence of the descent of man from some lower form', and the next chapter compared the 'mental faculties of man and the lower animals'. A specific section titled 'Belief in God, Spiritual Agencies, Superstitions', left little doubt as to Darwin's conviction that God was only one superstition among others. There was, he said, 'no evidence that man was aboriginally endowed with the ennobling belief in the existence of an Omnipotent God', and anthropologists had shown 'that numerous races have existed and still exist, who have no idea of one or more gods, and who have no words in their languages to express such an idea'.[97] He nonetheless took care to note immediately after this affirmation that this question 'is of course wholly distinct from that higher one, whether there exists a Creator and Ruler of the universe; and this has been answered in the affirmative by the highest intellects that have ever lived'.[98] For Darwin, feelings and emotions were also the result of evolution, and 'the difference in mind between man and the higher animals, great as it is, is certainly one of degree and not of kind'.[99] And if 'the ennobling belief in God is not universal with man', it is the social instincts, aided by 'active intellectual powers and the effects of habit', that 'naturally lead to the golden rule, "As ye would that men should do to you, do ye to them like wise"'; and this lies, Darwin concludes, 'at the foundation of morality'.[100]

In trying to rationally explain the emergence of morals and even of the idea of God, naturalists, archaeologists, anthropologists and historians thus completed the slow process of the exclusion of supernatural intervention from the field of legitimate scientific explanation. This application of scientific naturalism to the origins of man was in fact a total reversal of perspective. After having long sought to legislate the limits of science and impose the superiority of theology on natural philosophy, the official representatives of the Christian churches had been unable to prevent the naturalistic method from extending its mode of questioning to all phenomena. They finally had to adjust their discourse to accommodate the evidence of the historicity of nature and of humanity and admit that it was necessary to interpret sacred religious works in a more subtle way than the long dominant literal reading, so as to adapt them to the most recent scientific discoveries. Thus, what Kepler, Galileo and others had suggested at the beginning of the seventeenth century with regard to astronomical phenomena, would ultimately become acceptable to religious authorities with regard to human history and biblical writings only three centuries later. But this broad consensus of Christian elites, the fruit of a long and tortured history, would not, however, garner consent among truly fundamentalist religious groups who continue to believe in the literal truth of the stories they deemed dictated by God, and who therefore continue to exert pressure to limit the freedom of scientific research.

SCIENCE CENSORED

Censorships. *Defamatory qualifications given by theologians to persons or books that they happen not to like or that do not accord with their infallible ideas.*

Baron Holbach (1768)[1]

The long process of the autonomization of science from religion and the correlative exclusion of God as a legitimate explanation of natural phenomena, developments that we described in the previous chapter, did not come about without resistance. On the institutional level, the reactions of the Catholic Church took the form of censorship and the prohibition of works containing conclusions contrary to official or traditional religious interpretations of natural phenomena. As we will see in this chapter, these condemnations were not limited to the seventeenth century and the emblematic cases of Copernicus and Galileo, as many tend to believe, but continued across the centuries up to the abolition of the Index in the middle of the 1960s. Some of these condemnations were officially recorded in the Roman Index of Prohibited Books, while other Christian condemnations were subtler and less public so as not to fuel controversy. All aimed to remind scholars that they had to submit to religious authorities.

The self-censorship of Catholic scholars

Having heard of Galileo's condemnation only in November 1633, but still ignoring the details of the case that circulated as rumours through private correspondence, the philosopher René Descartes (1596–1650) confided to the French Minim friar, Marin Mersenne (1588–1648), that he would not, as expected, publish his book, *Traité du Monde* (*The World*), whose foundation was Copernican. It is worth quoting his letter at length:

> I inquired in Leiden and Amsterdam whether Galileo's World System was available, for I thought I'd heard that it was published in Italy last year. I was told that it had indeed been published but that all the copies had immediately been burnt at Rome, and that Galileo had been convicted and fined. I was so astonished at this that I almost decided to burn all my papers or at least to let no one see them. For I couldn't imagine that he – an Italian and, as I understand, in the good graces of the Pope – could have been made a criminal for any reason except than that he tried, as he no doubt did, to establish that the Earth moves. I know that some Cardinals had already censured this view, but I thought I'd heard it said that it was nevertheless being taught publicly even in Rome. I must admit that if the view is false then so are the foundations of my philosophy, for it clearly follows from them; and it's so closely interwoven in every part of my treatise that I couldn't remove it without damaging the whole work.

Descartes thus chose self-censorship because he 'didn't want to publish a discourse in which a single word would be disapproved of by the Church; so I preferred to suppress it rather than to publish it in a mutilated form'.[2]

A few months later, he wrote again to Mersenne to tell him that he was 'astonished that an ecclesiastic [it was Gilles Personne de Roberval] should dare to write about the Earth's motion, whatever excuses he may give'. For he thought (wrongly, in fact) that the members of the Inquisition 'seem to forbid even the use of this as a hypothesis in astronomy'. He added that since he did not see 'that this censure has been endorsed by the pope or by any Council, but only by a single congregation of the Cardinals of the Inquisition', he did not abandon all hope that his treatise could eventually 'see the light of day'.[3] It is possible to detect a certain irony in Descartes when he confessed to Mersenne, in December 1640, not to be 'sorry that the [Protestant] ministers are thundering against the movement of the Earth; perhaps this will encourage our own [Catholic] preachers to give it their approval!' A good Catholic having a 'firm faith in the Church's infallibility', Descartes had nonetheless no doubt about his own arguments on the movement of the Earth, and boasted not to be afraid 'that one truth may conflict with another!'[4] He may have been right on the epistemological level, but the question was political and related to the authority of the Church. He did not, moreover, give up on his appeals to his friends in Rome and asked Father Mersenne to write to Cardinal de Baigné, Apostolic Nuncio in Paris, to tell him that only the prohibition of the affirmation of the movement of the Earth prevented him from publishing his philosophy. Descartes also confided to a friend 'having given the order that a Cardinal who considered him a friend for many years, and who was one of the Cardinals of the Congregation that condemned Galileo, be consulted'.[5] According to his first biographer, Adrien Baillet, who published *La Vie de Monsieur Descartes* in 1691, the mysterious Cardinal was none other than Francesco Barberini (1597–1679), nephew of Pope Urban VIII and a member of the Holy Office. Although nothing confirms this intervention, it is clear that Descartes had done everything to obtain permission to speak of the movement of the Earth and to publish his *World*. Despite his optimism, his treatise appeared only posthumously in 1664. He had nonetheless published several extracts from it. As early as 1635 he confided to Mersenne that following Galileo's condemnation, he had 'detached [his Optics] completely from The World, and [was] planning to have it published separately quite soon', which he did in 1637 as an Appendix to his *Discourse on Method*.[6] A large part of his censured treatise also appeared in his *Principes de philosophie* (1644), in which he took great care not to assert the mobility of the Earth. This was ultimately a fruitless effort as his own works were put on the Index in 1663.[7]

In the years that followed Galileo's condemnation, confusion reigned as to the exact status of this decision by the Roman Inquisition. Even a Catholic priest like the astronomer Ismaël Boulliau (1605–1694) remained convinced that teaching the mobility of the Earth in no way constituted a gesture hostile to the Church. In December 1644, two years after the death of Galileo, he confided to Mersenne that it was possible that the prohibition to teach the movement of the Earth was 'something invoked only in Italy and not throughout Christendom, as no one has been notified by the Holy See; without doubt they have judged it inappropriate'.[8] This position, favourable to Copernicus and Galileo, was also, in private, that of the Canon Pierre Gassendi (1592–1655), the Parisian scholar and theologian. As early as 1629, he had written to Peiresc to say that almost all the scholars of the Netherlands were in favour of the movement of the Earth.[9]

Galileo's condemnation created a veritable shock wave amongst Catholics scholars who would, for the most part, refrain from openly discussing their adherence to the Copernican system. Thus, after having been openly Copernican, Gassendi lined up officially behind the Church and adopted Tycho Brahe's astronomical system that left the Earth stationary at the centre of the universe with the other planets orbiting around the Sun.[10] This astronomical system, preferred by the Jesuits, had not been discussed by Galileo in his *Dialogue*, as he considered it both artificial and arbitrary. Father Mersenne himself seemed ambivalent and hid behind the argument, common since the Middle Ages, of divine omnipotence. For God could have had 'good reason for making the sky rotate and keeping the Earth fixed'.[11] The Dutch Calvinist physicist Christiaan Huygens (1629–1695) wrote, in 1660, that whenever he spoke about Copernicus with Roman Catholics, they 'claim that they are not forced to accept the decrees against that theory'. He added that they were 'convinced that the immobility of the Earth must be defended with reasons rather than imposed by official documents'.[12]

Although many Catholic scientists expressed their scientific beliefs in private, they did otherwise in public. Since the official doctrine of Rome stated that the Copernican system could not be regarded as true, Catholic teaching institutions had to comply with this judgment. We find a clear expression of this obedience to the Church in Boulliau, who said that while he was convinced that the Pope would never intervene in matters that did not affect the faith, in the event that he did, he would reject the Copernican doctrine even though he believed it to be true.[13] Prudence was also clearly manifested by the once ardent Copernican Gassendi who, wanting to write to Galileo and knowing him to be under arrest, had contacted Peiresc in January 1634. The latter replied that he did not believe that Galileo was forbidden to receive letters but cautioned Gassendi 'to write them in terms so reserved and adjusted that there is a way to hear much of your intention without the literal sense being very precise'. Giving an example of such coded language, he concluded his letter saying that he was happy 'to hear of the work that Bernegger is doing in Strasbourg', an allusion to the Latin edition of Galileo's book, published two years after the original Italian edition had been put on the Roman Index of Prohibited Books.[14]

In July of the same year, Father Mersenne sent Peiresc his latest book, *Les Questions théologiques, physiques, morales et mathématiques*, noting that since the Sorbonne theologians had accused him of not refuting the movement of the Earth, he had written a new version in which he had 'removed all the questions that could offend them' and replaced them by other less problematic ones. This new edition, he added, would please his friend Giovanni Battista Doni in Rome.[15] In France, the theologians at the Faculty of Theology of the Sorbonne were responsible for the censorship of books. The Faculty had collated and published the first index of prohibited books in 1544, even before Rome had published its own.[16] Its theologians were supposed to 'oppose the emerging errors and elucidate the truth under the guidance of the Most Holy Mother Church and the Apostolic Holy See', as stated in a 1501 document from the Sorbonne's Faculty of Theology.[17]

The underlying reason for the reluctance of the Church to abandon the literal reading of the Bible was clearly expressed in the middle of the seventeenth century by the Jesuit astronomer Giovanni Battista Riccioli (1598–1671), a supporter of the immobility of the Earth and one who clearly understood the predictable consequences of an even partial abandonment of such literal and official reading. According to Riccioli, if the Church gave Copernicans the liberty to 'interpret scriptural texts and to elude ecclesiastic decrees', then one feared that this liberty 'would not be limited to astronomy and natural philosophy and could extend to the most holy dogmas'. It was therefore important, he declared in his *Almagestum Novum* of 1651, to maintain the rule that all sacred texts must be interpreted literally.[18] And it is worth noting in passing that Riccioli himself encountered problems with Roman censorship during the publication of his book as he complained in his letters to another Jesuit scholar, Athanasius Kircher.[19]

Censorship at the University of Louvain

The official condemnation by the Church of Rome of Copernican ideas was not trivial and targeted Catholic Christianity as a whole. The extent of Rome's power was clearly on display in the controversy that shook the University of Louvain in 1691-2, at the same time as Leibniz tried, in vain, to cancel the condemnation of the Copernican system as we saw in chapter 2.

At the beginning of 1691, Martin Etienne van Velden (1664–1724), a Professor in the Catholic University of Louvain, set out to have his students publicly support the veracity of the Copernican system, during the traditional *disputationes* of the Faculty of Arts, even though this was in 'disagreement with the decrees of the Sacred Congregations', as the Dean of the Faculty wrote to the Internuncio, the Pope's representative in Brussels.[20] The Dean, supported by the other teachers, ordered van Velden – under penalty of being forbidden to teach and thus losing his remuneration – to delete this thesis and replace it by another, less problematic, so as to avoid any confrontation with religious authorities. Deeply Catholic, the University of Louvain was well versed in the field of censorship, its Faculty of Theology having published in 1546 its own Index of Prohibited Books, thus mimicking the gesture made two years earlier by the Faculty in Paris. It published three catalogues containing a total of more than 700 condemnations, an enterprise that kept the theologians quite occupied with 'bad readings'. The Parisian Indexes had themselves censored a little more than 500 books.[21]

Aware that a professor of the faculty 'defended a thesis that does not express due reverence to the decrees of the Apostolic Holy See', the Internuncio quickly requested the Dean of the Faculty to keep him informed 'of the text of the thesis, the name of the teacher who supported it' and the measures that he intended to take in this regard. He diplomatically stated that he had no doubt that the faculty had done nothing, nor would do anything, that was not 'consistent with the expectation of His Holiness'.[22]

Despite the fact that the apostolic Internuncio was prepared to accept that the thesis be upheld if the author clearly indicated that he considered the Copernican system simply as a hypothesis, in accordance with the Decree of the Congregation of the Index of 1616, the Faculty refused any compromise. Faced with van Velden's questioning of the authority of the Rector, it reaffirmed its absolute autonomy in decision-making. Convinced he would never find justice within his institution, van Velden turned to the civil authorities even though the university did not recognize its competence in academic matters. He also contacted his compatriot Christiaan Huygens to keep him abreast of the situation. Denouncing the role played by the Faculty of Theology of the university in this censorship, van Velden noted that:

> if we do not propose an effective remedy for this evil, it will mean the end of Modern Philosophy here, because if I succumb (which in itself is little) no one will feel safe enough to dare mention Copernicus, Descartes, or your illustrious name or that of some new and learned philosopher.

He concluded his call for help by imploring Huygens 'in the name of all the lovers of Truth and Freedom' to kindly ask his brother Constantine, who was then close to the Governor, to promote his cause. This was his ultimate recourse in an attempt to bring back to reason those who stubbornly continued to prohibit the teaching of 'such an innocent and purely philosophical thesis'. Huygens did indeed write to his brother Constantine and asked him to 'say a word in favour of the petitioner so that he might be delivered from this persecution'. From Constantine's diary we learn that, armed with the letter from his brother, van Velden did meet him in August, and repeated that he felt 'persecuted for several theses that he had made in favour of the new philosophy. He wanted the King, or at least [him, Constantine], to speak or write to the Marquis of Gastanaga' who was then the Governor of the Netherlands.[23]

We do not know if this intervention had any effect, but the matter was finally resolved at the beginning of 1692. After many exchanges between the university, the government and the apostolic Internuncio, the Privy Council of the King confirmed the exclusive jurisdiction of the University on the issues raised by van Velden and invalidated the attempt by the local government to block any action by the Rector against the professor. The Rector had written directly to the King of Spain a week before the final verdict to remind him that not to dismiss van Velden's appeals to external authorities would only lead to the negation of the privileges granted by the King to the university and leave unpunished any future excesses committed by the subjects of the Rector.[24] While the Rector had seen his authority confirmed, the Internuncio asked him not to allow the case to fester but to conclude it as quickly as possible, forgiving his professor 'provided he showed due respect to the Holy See and submitted to the judgment of the Sovereign Pontiff the proposals contained in his theses or those he would later insert in them'.[25] Defeated, van Velden agreed to submit and was condemned to pay a portion of the legal fees incurred by the university. He also signed the following statement that recalls the one imposed on Galileo:

> I published and defended at the University on the 11th and 12th of July 1691, philosophical arguments, some of which displeased, especially those concerning the system of Copernicus and the Decree of his Holiness, our Lord Pope Urban VIII; similarly, those on how to teach philosophy and examine the students of the University and on the manners to discuss that are generally in use in all schools; finally concerning those who oppose the doctrine of René Descartes. So, I declare that it was not my intention to offend the Holy See, the Magnificent Lord, the University or my colleagues; I regret now that the things in my thesis that have crept in and that may have given occasion to offend. I will from now on be more careful so that similar things do not happen again. I will also make a similar statement to the illustrious Apostolic Internuncio, regarding things that could have offended the Holy See.[26]

Ceding to the pressures of his surrounding, he no longer spoke openly of the movement of the Earth. In exchange for this submission, he rose within the ranks of the local elite.[27]

Little known, this controversy clearly illustrates the logic of the doctrinal control provided by Rome within the Catholic world. The

Vatican authorities were constantly informed by the vast network of Nuncios and leaders of Catholic universities. While van Velden's personality no doubt played a role in this story, we must not lose sight of the fact that this controversy, which opposed the three institutions seeking to assert their jurisdiction and to impose their authority (the university, the local civil government and Rome), was only possible because Rome had prohibited the teachings of the Copernican system as describing reality. The Catholic universities had to obey the orders of the Index and of the Inquisition. Writing to a representative of the King's Privy Council responsible for this matter in Brussels, van Velden even admitted that he was 'astonished that for such a trifle thesis of no importance so much noise had been made' and was even 'ashamed to have to run so often to Brussels and to annoy everyone for such a subject'.[28]

Vexing atoms for the Church

Even though celestial physics preoccupied the guardians of religious orthodoxy, it did not prevent them from keeping an eye on terrestrial physics. At the beginning of the seventeenth century, the renewed interest in the ancient doctrine of atoms troubled philosophers and Catholic theologians able to draw the logical consequences of a philosophy that plainly veered in the direction of atheism and materialism.

The idea that the world is made up of atoms in random movement in a vacuum dates back to the Greek philosophers Leucippus and Democritus and was popularized by the Latin poet Lucretius in his book *On the Nature of Things*, the text of which was rediscovered in 1417. The revival of atomism fit into an intellectual movement against the dominant scholasticism based on the theories of Aristotle, which had been the basis of the teaching of philosophy and Christian theology since the Middle Ages.[29] According to the Aristotelian theory of matter, indivisible atoms and the vacuum could not exist: all matter was composed of primary (substance) and secondary (accidents) qualities. This dual aspect of matter coincidentally allowed Catholic theologians to rationally explain the miracle of transubstantiation, a fundamental dogma of the Roman Church. Indeed, when the priest pronounces the liturgy and blesses the bread and the wine, a miraculous operation occurs which, according to Catholic orthodoxy, transforms these substances respectively into the Body and Blood of Christ while leaving their accidents, that is

101

their outside appearances, intact, because the communicants obviously taste the bread and the wine. The miracle thus transforms the substance of the bread and wine but leaves their 'accidents' (here the sensations) unchanged. From the atomistic point of view, this is impossible because transforming the substance necessarily alters the sensations, the atoms having only primary and no secondary qualities. In this system, sensations are the effects of the interaction of the atoms with the senses. It is thus understandable that any work based on atomism ran the risk of being attacked for religious reasons and not just for philosophical motives.

Galileo also claimed to be an atomist in his 1623 essay, *Il Saggiatore* (*The Assayer*). The book was also denounced to the Inquisition by an anonymous author who noted that 'if this philosophy of qualities is admitted to be true, it seems to [him] there follows a great difficulty in regard to the existence of the qualities of bread and wine which in the Holy Sacrament are separated from their own substance'.[30] Galileo's theory, which incorporated Democritus, seemed to him to 'go against the common view of theologians' and to be 'repugnant to the truth of the Sacred Councils'.[31] The Jesuit Orazio Grassi would say as much in a treatise published in 1626:

> In the host, it is commonly affirmed, the sensible species (heat, taste, and so on) persist. Galileo, on the contrary, says that heat and taste, outside of him who perceives them, and hence also in the host, are simple names; that is, they are nothing. One must therefore infer, from what Galileo says, that heat and taste do not subsist in the host. The mere thought of such a possibility is frightening. The soul experiences horror at the very thought.[32]

For unknown reasons, the Inquisition did not follow up on the anonymous whistleblower's complaint.[33] At the same time, however, the Faculty of Theology of the Sorbonne, with the support of the Parliament of Paris, prohibited public discussion of fourteen 'theses' that opposed the philosophy of Aristotle, including those promoting atomism and gave the authors of the theses twenty-four hours to leave Paris.[34] In 1632, an internal regulation of the Jesuits order also prohibited the teaching of atomistic ideas in their colleges and schools.[35] Following the work of the priest Pierre Gassendi, which proposed an interpretation of atomism acceptable for Catholics – in much the same way as Thomas Aquinas had Christianized Aristotle – and the dissemination of Descartes' philosophy in the middle of the century, the question of the conflict between atomism and theology became much discussed among philosophers. In 1658, a philosophical essay

102

printed in Rome reminded its readers that the 'opinion denying generally that accidents are separate from corpuscles presents a difficulty from the perspective of faith and of the Eucharistic mystery'.[36] And even though Descartes clearly denied believing in atoms and claimed instead that matter was infinitely divisible, his concept of 'corpuscles' was most often understood as consistent with atomistic theses and associated with the denial of the reality of accidents.[37] His works, including his *Principles of Philosophy* that contains his physics, had, moreover, been placed on the Index – subject to some corrections as had been the case for Copernicus – in 1663.[38] The hostility to atomism would ultimately be reflected in a decree of the Inquisition in 1673 ordering the local inquisitors to refuse the 'imprimatur' to works supporting this doctrine.[39] But unlike heliocentrism, the Church never went so far as to officially condemn the doctrine by placing already published books on the Index and sought merely to curb its dissemination. Even in the middle of the eighteenth century, the Congregation of the Index bothered the Neapolitan theologian Antonio Genovesi, author of a metaphysical treatise that was too eclectic for the Church, by sending him a list of thirty-four corrections to be made, including one on atomism because it was presented as compatible with Christian doctrine.[40]

If, in the end, atomism managed to impose itself in chemistry and physics at the beginning of the twentieth century, the question of the nature of substances remained problematic at the theological level. As late (or as recently . . .) as 1950, in his Encyclical *Humani Generis*, Pius XII still felt obliged to recall that too many errors 'have crept in among certain of Our sons who are deceived by imprudent zeal for souls or by false science'. He felt compelled 'with grief to repeat once again truths already well known, and to point out with solicitude clear errors and dangers of error'. Among these truths worth repeating we find the scholastic concept of substance, which continued to be necessary because the atomic theory remained of course incompatible with the dogma of transubstantiation. The Pope therefore called to order those who 'even say that the doctrine of transubstantiation, based on an antiquated philosophic notion of substance, should be so modified that the real presence of Christ in the Holy Eucharist be reduced to a kind of symbolism, whereby the consecrated species would be merely efficacious signs of the spiritual presence of Christ and of His intimate union with the faithful members of His Mystical Body'.[41] In sum, the adaptation of religious conceptions to even those scientific theories with the most solid foundations remain difficult for Roman authorities.

The plurality of worlds

Both atomistic theses and Copernican cosmology entailed, for different reasons, the existence of a multitude of inhabited worlds. By making the Earth a planet like the others and the Sun a star like other stars, the Copernican system quite naturally suggested that there was no reason to believe that our planet was the only one inhabited by intelligent beings created by God. Atomists of the seventeenth and eighteenth centuries came to the same conclusion in arguing that space being infinite and our world being but the result of fortuitous collisions between atoms that were eternal and infinite in number, there should necessarily exist other inhabited worlds.[42]

For Catholic theologians, the plurality of worlds raised many problems since the Holy Scriptures spoke only of the sins of Adam and Eve in an *earthly paradise* and the salvation of humanity on this planet without any mention of any other inhabited world. To assert or even speculate about the existence of peoples living on the Moon or elsewhere in the universe could therefore only be foolhardy if not heretical. The atomist Gassendi considered the question insoluble. The power of God being infinite, He would have been able to create other worlds if He had wanted to, but as we have been unable to observe them, it was impossible to know. In addition, the idea that our world was unique seemed to him 'most in conformity with the principles of the Sacred Faith and religion' because everything that was said in the Sacred Scriptures 'with regard to the origins of things indicate a single world'.[43]

Despite the theological uncertainty of the plurality of worlds, religious authorities preferred to avoid controversy by controlling publication. Thus, Fontenelle's (1657–1757) *Entretiens sur la pluralité des mondes*, published in 1686, was quickly denounced – remember that it was the duty of any Catholic to denounce suspect books – and the Congregation of the Index was swift to condemn the book, the decree of censorship having been issued promptly on 22 September 1687. According to the Consultor responsible for assessing the book, the question of the plurality of the worlds was theologically complex and difficult, but the book was 'reckless and dangerous for the salvation of souls' and therefore had to be prohibited.[44]

The Congregation of the Index was slower to prohibit another book on the plurality of worlds, that of John Wilkins (1614–1672), probably because it was written in English by an Anglican bishop, moreover a mathematician and future secretary of the Royal Society of London. Entitled *A Discovery of A World in the Moone or,*

104

A Discourse Tending to Prove that It Is Probable There May Be Another Habitable World in That Planet, it had been published in English in 1638, only five years after the condemnation of Galileo. Republished several times under varying titles, and translated into French in 1655 under the title *Le Monde dans la Lune*, it was finally put on the Index in 1701. The book contained many heretical statements brought together as 'Propositions that are insisted on in this discourse' at the beginning of the book. Thus, the second Proposition of the first part of the book stipulates that 'a plurality of worlds doth not contradict any principle of reason or faith'. A few decades earlier, Giordano Bruno had also been condemned for having affirmed the plurality of inhabited worlds and was burned alive in a public square in Rome for having refused to recant his convictions, including his rejection of Mary's virginity.

Wilkins also criticized the literal reading of the Bible and promptly affirmed in the second proposal of the second book, 'that the places of Scripture, which seem to intimate the diurnal motion of the Sun or Heavens, [are] fairly capable of another interpretation'. Proposal 5 added 'that the words of Scripture, in their proper and strict construction, do not anywhere affirm the immobility of the Earth'. In the tradition of Kepler and Galileo, his proposition 3 affirmed 'that the Holy Ghost in many places of Scripture, does plainly conform his expressions to the error of our conceits, and does not speak of sundry things as they are in themselves, but as they appear to us'. In passing, he slighted theologians by affirming (in proposal 4 of the second book), 'that diverse learned men have fallen into great absurdities, whilst they have looked for the grounds of philosophy from the words of Scripture'.[45] Like many scholars before him, he cited the Greek philosopher Alcinous to the effect that it 'behoves every one in the search of truth, always to preserve a philosophical liberty'.[46]

Given that astronomers currently seek intelligent life in other solar systems, theologians seem to have become accustomed to the idea that – after all – far from being heretical, the plurality of worlds could even prove to be more consistent with the exercise of the infinite power of God than its unicity. A Protestant theologian recently proposed, in a paper published in the *Philosophical Transactions* of the Royal Society of London, generally regarded as a 'prestigious' scientific journal, that the discovery of intelligent extraterrestrial life would only confirm the idea that the entire universe, and not only the Earth, 'will be seen as the gift of a loving and gracious God'; and these creatures could also be saved from their sins . . . [47]

No astronomy for ladies . . .

The passage of time and especially the advancement of the physical sciences made the prohibitions on the presentation of the Copernican system as consistent with reality increasingly untenable and just as increasingly embarrassing for Catholic scholars. But this did not prevent the Congregation of the Index from continuing its work. In 1737, for example, it prohibited the reading of the popular science work *Il Newtonianismo per le dame*, published anonymously and written by a friend of Voltaire, Francesco Algarotti (1712–1764). A bestseller immediately translated into French and English, it presented a Newtonian physics that confirmed the Copernican system as revised by Kepler. For this reason, the book was put on the Index in 1739.

At the same time, a Latin edition of Newton's fundamental work, the *Principia Mathematica*, with notes by the mathematicians and Minim friars François Jacquier (1711–1788) and Thomas Le Sueur (1697–1770), appeared in Geneva. In this book, first published in 1687, Newton formulated the theory of gravitational attraction to physically explain the movement of the Earth around the sun, thus providing a physical foundation for the Copernican system. The book had never been put on the Index, perhaps because it had already been implicitly condemned by the Decree of 1616 prohibiting 'any writing' promoting the movement of the Earth. As Book 3 dealt with the system of the world, the two religious translators felt obliged to point out that 'in this third volume, Newton adopts the hypothesis of a moving Earth', even though they themselves accepted 'the decrees of the popes against the movement of the Earth'.[48] By employing the term *hypothesis* (instead of *thesis*) to explain Newton's reasoning, the authors managed to avoid trouble with the Inquisition and the Index.

Almost a hundred years later, in 1830, the popular science book *Astronomy for Ladies*, published forty-five years earlier by the great astronomer and professor at the Collège de France, Joseph Jérôme Lefrançois de Lalande (1732–1807), was, in turn, added to the list of banned books by the Catholic Church.[49] Perhaps he, too, had been wrong to explain, like so many others, that the language of the Bible was that of common sense and not of science, and especially to have asserted, in reference to the famous passage of the Bible where Joshua stops the Sun, that it was 'very strange that one could pretend that Joshua used the language of philosophers, unknown to his country and his time'. The astronomer even dared to affirm that it was 'somewhat stupid to pretend that an army general like Joshua, who, when he had to show to his soldiers the glory and power of God by a victory, had to give them an astronomy lesson and, abandoning the language they could understand, tell the Earth to stop', instead of the usual common sense affirmation that the Sun rises and sets.[50]

A 'hypothetical' *Natural History*

The French naturalist Buffon was also forced to use the subterfuge of a 'hypothesis' to free himself from the troubles caused by the Sorbonne theologians following the publication of his *Natural History* in 1749. Offering an interpretation of the history of the Earth and the planets clearly incompatible with the biblical narrative, Buffon saw his ideas quickly denounced by the more conservative minds of the Sorbonne's Faculty of Theology, the guarantor, in France, of the authority of the Catholic Church.[51] He nonetheless had hoped it would not be put on the Index for he had done everything 'to avoid the theological annoyances that [he] fear[ed] much more than the critics of physicists and mathematicians'.[52]

But the theologians could not remain silent when faced with his comments and, on 15 January 1751, Buffon received a letter from Trustees of the Faculty of Theology of Paris who let him know that his *Natural History* was one of the works chosen to be 'examined and censured' as it contained 'principles and maxims not conforming to those of religion'. In the great tradition dating back to the Middle Ages, the Faculty identified fourteen 'propositions' extracted from the first four volumes of the book that 'seemed reprehensible' to the representatives of religious orthodoxy.[53] The censors had been careful to identify extracts in both the large format (in Quarto) edition and the popular edition (in-12), the existence of the latter indicating the widespread readership of the naturalist's writings.

While the majority of the extracts identified by the theologians focused on philosophical issues relating to the concept of truth and the nature of the soul, the first four related to the natural history of the solar system which presented the planets as having been formed from the Sun:

I. It is the waters of the sea that have produced the mountains, the valleys of the Earth ... it is the waters of the sky that, levelling everything, will one day return this land to the sea, which upon being dispossessed will leave new continents, similar to those on which we now live, to be discovered (*edit. in-4° tome I, p. 124; in-12, tome I, p. 181*).

107

II. Can we not imagine . . . that a comet falling on the surface of the Sun might have moved this star, and that she might have separated a few small parts of it to which an impetus was transmitted . . . so that the planets might have once belonged to the body of the sun and would have since become detached (*edit in-4° p. 133; in-12, p. 193*).

III. Let us see in what state they (the planets, & especially the Earth) are found, after having been separated from the mass of the Sun (*edit in-4° p. 143; in-12, p. 208*).

IV. The Sun will probably die off . . . from lack of combustible material . . . the Earth borne out of the Sun was therefore extremely hot and in a state of liquefaction (*edit. in-4°, p. 149 in-12, p. 217*).[54]

As Buffon was the Director of the *Jardin du roi* and a prominent member of the Paris Academy of Sciences, the theologians had to find a way to both accommodate the conservatives who were lobbying to put the book on the Index and to work out a solution acceptable to the Court, the book having been printed at the expense of the King and become a popular bestseller.[55] Buffon agreed to publish a reply in which he explained that he never had 'any intention of contradicting the text of Scripture', that he firmly believed everything in it that concerned the Creation of the world and that he abandoned anything in his book, that 'touches on the formation of the Earth, and in general everything that might be contrary to the narrative of Moses'. He affirmed that he had presented his 'hypothesis on the formation of the planets only as a purely philosophic supposition'.[56] Despite his central position in France's scientific community, Buffon was thus forced to accept Bellarmine's Catholic epistemology that allowed astronomers of Galileo's time to make assumptions on the functioning of the universe without asserting that their model truly described reality, a skilful way to save the 'truth' of the Scriptures and to avoid official condemnation by the Church. Buffon also agreed to publish the theologians' letters and his response in volume IV of his *Natural History* that appeared in 1753. Some years later, in 1760, he confessed to a friend the opportunistic nature of his submission to the religious authorities.[57]

Thanks to this compromise, Buffon considered the affair closed, as he confided to Abbot Le Blanc in 1751 and deplored the fact that Montesquieu, who had published *The Spirit of the Laws* in 1748, was still grappling with the theologians – the book was put on the Index in 1751. He was even surprised to get unexpected positive comments from members of the Faculty of Theology.[58]

Thirty years later, however, his book on *Les Époques de la nature* was denounced to the authorities on the eve of its publication in 1779. In the book, he had explained the geological history of the

Earth as a series of periods suggesting that the 'days' of the biblical story were in fact 'eras' of great duration, a proposition contrary to the literal reading of the Bible. But Buffon was then old, renowned and well-received in Court circles, so the theologians of the Sorbonne were loath to condemn him for fear of displeasing the King. The incident therefore ended quickly. Confiding to a friend shortly after the publication of the book, and speculating on the identity of his denouncer, Buffon wrote: 'I do not think this case will lead to anything beyond having unfortunately to hear about it and perhaps to provide an explanation as stupid and absurd as the first one they made me sign thirty years ago'.[59]

The condemnation of materialism and the critical history of religion

In the course of the nineteenth century, it was often the scent of materialism that pushed the Consultors of the Congregation of the Index to recommend the prohibition of scientific books. In 1834, after reading an otherwise technical book by the French physician, chemist and politician François-Vincent Raspail (1794–1878), *Nouveau système de chimie organique fondé sur des méthodes nouvelles d'observation*, published the previous year, a consultor wrote that 'the more it is correct [scientifically] the more it is dangerous because readers will finally be misled because they admire the rigor of its physical and chemical arguments and [. . .] they will take as truth the mistakes he teaches'. The chemist had in effect 'the defect of all those who profess his science' and 'ignores the metaphysical essence of these beings through which their substantial difference is constituted'.[60] The book, which nonetheless went through several revised editions and was translated into Italian and German, was actually put on the Index by a decree of 1834. But rather than waiting for 200 years like Copernicus and Galileo to be removed from the list of prohibited books, Raspail's book fared better and recovered its 'freedom' in 1900, when the Roman authorities decided to do a little housekeeping and to delist the books whose condemnation had become anachronistic or obsolete.[61]

It was probably this same 'materialism' that explains that the work of the physician and professor of pathology François Joseph Victor Broussais (1772–1838), entitled *De l'irritation et de la folie* (1828), was also prohibited by a decree of 1830. This was in some respects the logical consequence of the 1819 condemnation of

Cabanis' (1756–1808) *Rapports du physique et du moral de l'homme*, published in 1802.

Rome's delay in condemning 'materialist' science is often explained by the absence of an Italian version of the book to be censured, the censors being known for their ignorance of foreign languages.[62] Lalande's book, *Astronomy for Ladies*, cited above, went through many editions after its release in 1795 but was not translated into Italian until 1828 and was finally condemned only in 1830.[63] Similarly, the six volumes of the German physiologist Karl Friedrich Burdach (1776–1847), Professor of Anatomy at the University of Köningsberg, *Die Physiologie als Erfahrungswissenschaft* (Physiology as an Empirical Science) was put on the Index in 1851, only after the first five volumes had been translated into Italian.[64] The same is true for the largely speculative and far too materialist for the Church's taste, *Zoomania, or the Laws of Organic Life*, by Erasmus Darwin (1731–1802), Charles Darwin's grandfather.[65] Published in English in 1794, it was put on the Index in 1817, seven years after its translation into French and more than ten years after its Italian translation. Among the 'materialist' works prohibited to Catholic readers, we should mention the famous – and always informative – *Course of Positive Philosophy* by Auguste Comte (1798–1857), published in six volumes between 1830 and 1842 and put on the Index of Prohibited Books in 1864. The wonderful *Sketch for a Historical Picture of the Progress of the Human Mind*, written in haste by Condorcet while hiding from supporters of the revolutionary terror, and published immediately after his death in 1794, would also be added to the Index in 1827.

Even if most Protestant sects did not have centralized institutional structures comparable to those of the Catholic Church, in the nineteenth century the Anglican Church still controlled educational institutions and was thus able to exert pressure on authors it considered 'materialist'. The professor of physiology William Lawrence (1783–1867) learned this at his own expense when he published in 1819 his *Lectures on Comparative Anatomy, Physiology, Zoology, and the Natural History of Man*, much too materialist for the minds of the Anglican elites. His treatment of the 'natural history of man' argued that the mental processes that determine thoughts and consciousness were only a function of the brain. His remarks were immediately denounced in court as blasphemous and he was forced to abandon his chair at the Royal College of Surgeons. To calm things down, he withdrew his book from circulation. This submission to the Anglican authorities allowed him to resume his position a few years later, and he ended his career at the top of the social ladder as Surgeon-General to Queen Victoria.[66]

As we saw in the previous chapter, the critical history of religion, developed as a scientific conception of historical research and proposed in Germany by David Friedrich Strauss and promoted with even greater impact by Ernest Renan, did not go unnoticed by the censors of the Congregation of the Index. Strauss' *Life of Jesus* was prohibited as early as 1836, only one year after its release, and that of Renan suffered the same fate in the year of its publication, in 1863. A prolific author, Renan saw twenty of his books denounced by Rome and placed on the Index of Prohibited Books between 1859 and 1894, including his *Studies of Religious History*, published in 1857 and condemned by Rome in 1859. Renan's point of view reflected a radical rationalist current of thought quickly condemned by Rome in its *Syllabus of Errors* issued in 1864 by Pope Pius IX. Bearing on a variety of intellectual, political and social issues, it denounced eighty propositions deemed to be erroneous. The first two sections dealt with pantheism, naturalism and rationalism (absolute and moderate) and denounced fourteen propositions, most of which affirming the complete autonomy of science in relation to religious dogmas (Table 4.1).[67]

Proposition 14 took direct aim at exegesis and the critical history of religion, of which Renan was the most ardent promoter. The document ended with a condemnation of modern liberalism and of the proposal that 'The Roman Pontiff can, and ought to, reconcile himself, and come to terms with progress, liberalism and modern civilization', which nicely summarizes the spirit of the document.

A good Catholic, and less controversial than his compatriot but just as much a supporter of critical history, the Assyriologist François Lenormant (1837–1883) was aware of the religious problems raised by the progress of his science. Thus, the discovery, in the second half of the nineteenth century, of cuneiform texts much older than the Bible (including a part of the Epic of Gilgamesh), and recording legends identical to those contained in the first eleven chapters of *Genesis* posed difficult interpretative problems for Catholic theologians.[68] But for Lenormant, these facts were indisputable and could not be set aside to please religious authorities. In 1880, he set out his views in his book *Les Origines de l'histoire d'après la Bible et les traditions des peuples orientaux*, translated into English two years later. In a long preface, he reminded his readers that, as a historian, his book 'was composed without any other purpose than that of sincere and conscientious search after scientific truth'. However, he admitted that

Table 4.1

Some of the propositions condemned
by the Pope in 1864 (*Syllabus of Errors*)

I. Pantheism, naturalism and absolute rationalism

2. All action of God upon man and the world is to be denied.

3. Human reason, without any reference whatsoever to God, is the sole arbiter of truth and falsehood, and of good and evil; it is law to itself, and suffices, by its natural force, to secure the welfare of men and of nations.

4. All the truths of religion proceed from the innate strength of human reason; hence reason is the ultimate standard by which man can and ought to arrive at the knowledge of all truths of every kind.

6. The faith of Christ is in opposition to human reason and divine revelation not only is not useful, but is even hurtful to the perfection of man.

7. The prophecies and miracles set forth and recorded in the Sacred Scriptures are the fiction of poets, and the mysteries of the Christian faith the result of philosophical investigations. In the books of the Old and the New Testament there are contained mythical inventions, and Jesus Christ is Himself a myth.

II. Moderate Rationalism

8. As human reason is placed on a level with religion itself, so theological must be treated in the same manner as philosophical sciences.

9. All the dogmas of the Christian religion are indiscriminately the object of natural science or philosophy, and human reason, enlightened solely in an historical way, is able, by its own natural strength and principles, to attain to the true science of even the most abstruse dogmas; provided only that such dogmas be proposed to reason itself as its object.

10. As the philosopher is one thing, and philosophy another, so it is the right and duty of the philosopher to subject himself to the authority which he shall have proved to be true; but philosophy neither can nor ought to submit to any such authority.

11. The Church not only ought never to pass judgment on philosophy, but ought to tolerate the errors of philosophy, leaving it to correct itself.

12. The decrees of the Apostolic See and of the Roman congregations impede the true progress of science.

13. The method and principles by which the old scholastic doctors cultivated theology are no longer suitable to the demands of our times and to the progress of the sciences.

14. Philosophy is to be treated without taking any account of supernatural revelation.

it 'directly touches questions of the utmost gravity and of a particularly delicate nature', which obliged him to explain to the reader the spirit in which he had addressed these subjects. Claiming to be proud to be Christian, he immediately added that he was also 'a scholar, and as such [he did] not recognize both a Christian science and a science of free thought. [He] acknowledge[s] one science only, needing no qualifying epithet, which leaves theological questions on one side, as foreign to its domain, and accepts all investigators, working in good faith, whatever their religious convictions, as equally its servants'.[69] Like many other believers before him, he claimed that he had never 'come face to face with a genuine conflict between science and religion' for 'the two domains are absolutely distinct and not exposed to collision. There can be no quarrel between them, unless one encroach improperly upon the territory of the other. Their truths are of a different order'. And should any 'apparent antinomy between science and religion' appear, he was 'certain beforehand that a day will come when they will attain a harmony which [he] should not have been skilful enough to discover'.[70]

With regard to the status of the Scriptures, Lenormant firmly believed in their divine inspiration concerning faith and morals and subscribed 'with absolute submission to the doctrinal decisions of the Church in this respect'. For the rest, he accepted Galileo's point of view, citing the same sentence used by the astronomer to defend the autonomy of astronomy with regard to theology, reminding that the intent of the Holy Scripture was 'to teach us how to go to heaven, and not how the heavens go'.[71] Lenormant concluded optimistically that:

The submission of the Christian to the authority of the Church, in all that relates to those teachings of faith and morals to be drawn from the Books of the Bible, does not at all interfere with the entire liberty of the scholar, when the question comes up of deciding the character of the narratives, the interpretation to be accorded to them from the historical stand-point, their degree of originality, or the manner in which they are connected with the traditions found among other peoples, who were destitute of the help of divine inspiration, and lastly, the date and mode of composition of the various writings comprised in the scriptural canon. Here scientific criticism resumes all its rights. It is quite justified in freely approaching these various questions, and nothing stands in the way of its taking its position upon the ground of pure science, which demands the consideration of the Bible under the same conditions as any other book of antiquity, examining it from the same standpoints and applying to it the same critical methods. And we need fear no diminution of the real authority of our Sacred Books from examination and discussion of this nature, provided that it be made in a truly impartial spirit, as free from hostile prejudice as from narrow timidity.[72]

This argument in favour of a distinction between different types of statements contained in the Bible was not considered convincing at the time of Galileo by the Roman authorities. Applied to biblical texts, the reasoning remained still too bold at the end of the nineteenth century, and Lenormant's book was immediately attacked and finally put on the Index in 1887, a few years after his death.[73]

After decades of polemics, the thesis according to which not all biblical statements had the status of revealed truth, was officially rejected by the Church in 1893, in the Encyclical *Providentissimus Deus*. Leo XIII thus lay down the limits of legitimate positions on biblical exegesis and confirmed that interpreters should generally stay close to the text. There was a recognition that 'copyists have made mistakes in the text of the Bible; this question, when it arises, should be carefully considered on its merits, and the fact not too easily admitted, but only in those passages where the proof is clear. It may also happen that the sense of a passage remains ambiguous, and in this case good hermeneutical methods will greatly assist in clearing up the obscurity'. That said, the Pope reaffirmed that 'it is absolutely wrong and forbidden, either to narrow inspiration to certain parts only of Holy Scripture, or to admit that the sacred writer has erred'.[74]

The work of critical history continued nonetheless on its course and the so-called modernist movement within the Catholic Church would finally be condemned in 1907 by Pope Pius X in his Encyclical *Pascendi*. The latter had been preceded a few weeks earlier by the decree *Lamentabili* that, still within the tradition that went back to the Middle Ages, forbade sixty-five 'propositions' deemed incorrect from the point of view of faith. This decree, a sort of update of the *Syllabus of Errors* of 1864, was published so that these errors that 'are being daily spread among the faithful' and could thus 'captivate the faithful's minds and corrupt the purity of their faith' be clearly condemned. So, following 'a very diligent investigation and consultation with the Reverend Consultors, the Most Eminent and Reverend Lord Cardinals, the General Inquisitors in matters of faith and morals have judged' that sixty-five propositions had to be 'condemned and proscribed'.[75] A large number of them related directly to the historical criticism of the Scriptures that called into question the monopoly of theologians and Roman Church in the interpretation of the texts (Table 4.2).

The rejection of these proposals reflected a very conservative view of the relationship between faith and reason, one that could only lead

---

Table 4.2

Some of the propositions condemned
by the Pope in the Decree *Lamentabili* of 1907

1. The ecclesiastical law which prescribes that books concerning the Divine Scriptures are subject to previous examination does not apply to critical scholars and students of scientific exegesis of the Old and New Testament.

2. The Church's interpretation of the Sacred Books is by no means to be rejected; nevertheless, it is subject to the more accurate judgment and correction of the exegetes.

3. From the ecclesiastical judgments and censures passed against free and more scientific exegesis, one can conclude that the Faith the Church proposes contradicts history and that Catholic teaching cannot really be reconciled with the true origins of the Christian religion.

4. Even by dogmatic definitions the Church's magisterium cannot determine the genuine sense of the Sacred Scriptures.

5. Since the deposit of Faith contains only revealed truths, the Church has no right to pass judgment on the assertions of the human sciences.

11. Divine inspiration does not extend to all of Sacred Scriptures so that it renders its parts, each and every one, free from every error.

12. If he wishes to apply himself usefully to biblical studies, the exegete must first put aside all preconceived opinions about the supernatural origin of Sacred Scripture and interpret it the same as any other merely human document.

13. The Evangelists themselves, as well as the Christians of the second and third generation, artificially arranged the evangelical parables. In such a way they explained the scanty fruit of the preaching of Christ among the Jews.

14. In many narrations the Evangelists recorded, not so much things that are true, as things which, even though false, they judged to be more profitable for their readers.

15. Until the time the canon was defined and constituted, the Gospels were increased by additions and corrections. Therefore there remained in them only a faint and uncertain trace of the doctrine of Christ.

23. Opposition may, and actually does, exist between the facts narrated in Sacred Scripture and the Church's dogmas, which rest on them. Thus the critic may reject as false facts the Church holds as most certain.

24. The exegete who constructs premises from which it follows that dogmas are historically false or doubtful is not to be reproved as long as he does not directly deny the dogmas themselves.

115

Table 4.2. (continued)

32. It is impossible to reconcile the natural sense of the Gospel texts with the sense taught by our theologians concerning the conscience and the infallible knowledge of Jesus Christ.

34. The critics can ascribe to Christ a knowledge without limits only on a hypothesis which cannot be historically conceived and which is repugnant to the moral sense. That hypothesis is that Christ as man possessed the knowledge of God and yet was unwilling to communicate the knowledge of a great many things to His disciples and posterity.

36. The Resurrection of the Saviour is not properly a fact of the historical order. It is a fact of merely the supernatural order (neither demonstrated nor demonstrable), which the Christian conscience gradually derived from other facts.

to the condemnation of the promoters of this 'modernism', at the centre of which stood the priest and theologian Alfred Loisy (1857–1940). Indeed, several of the condemned statements were drawn directly from his works. A historian of religion and a student of Renan at the Collège de France at the beginning of the 1880s, Loisy had first been forced to resign from his position as a professor of biblical exegesis at the Catholic Institute of Paris in 1893 due to his overly 'modern' ideas, in favour of an enlightened Catholicism and of the strict separation between historical exegesis and faith.[76] Like Renan and Lenormant, he was opposed to the very idea of a 'Catholic science'. Considered a new Renan by his enemies, Loisy observed at the beginning of the century that 'the Church is now obliged to endure the scientific movement that occurs outside her, but she tries to keep it where it is, that is outside her, and jealously guard her own science – called, without laughing, Catholic science – against any profane contact'.[77]

In 1903, five of his works were put on the Index. But not having 'abjured his errors' and having rather 'confirmed them with obstinacy in new writings and in letters to his superiors', he was finally excommunicated in 1908, by a decree of the Supreme Congregation of the Holy Office that, 'on the express mandate of our Holy Father Pius X, pronounced the sentence of major excommunication'. In the Catholic world, Alfred Loisy thus became a person who 'must be avoided by all'.[78]

The reaction of his lay supporters was immediate: the next year, Loisy was elected Professor at the prestigious Collège de France, to the Chair of the History of Religions. He took advantage of his inaugural lecture to reaffirm that:

like any other form or manifestation of thought, activity, of human life, religions can be a matter for science. Nothing prevents the application to all religious facts, to all types of documents and testimony that concern them, of the methods of observation and criticism employed in other sciences that have humanity as their object, especially in History.[79]

Replying to the theologians who opposed the complete autonomy of the history of religion, he observed that 'the science of religions cannot serve any particular theology, and specific theologies are incompatible with the science of religions' because 'they would impose on that science conclusions that often contradict its results and would always hinder research'.[80] In short, 'the science of religion is not a religion'.[81]

Recuperated by a secular, republican institution, Loisy published numerous books over the following decades, most of which were condemned by Rome: two were put on the Index in 1932 and eight more in 1938.[82] Here, as in other areas of the relationship between science and faith, the Roman Church would eventually retreat, and the Encyclical *Divino Afflante Spiritu*, published by Pius XII in 1943, would in part legitimate what had previously been condemned. At the beginning of the 1960s, the Vatican II Council took a step further in the acceptance of the historicity of the biblical texts.[83]

Evolutionists' retractions

Most of the books published during the last twenty years on the relationship between science and religion take great pains to point out that the works of Darwin, above all *On the Origin of Species*, were never formally placed on the Index. They often forget to add, however, that other books by evolutionists were prohibited by religious authorities. In fact, as early as 1860, Darwin himself was explicitly condemned – but not named because the decrees rarely named their targets – by the German Catholic episcopate meeting in regional Council in Cologne.[84] The German translation of Darwin's book having barely been published, the Bishops declared 'quite contrary to the Holy Scripture the opinion of those who are not ashamed to say that the body of man is the result of the spontaneous transformation of an imperfect nature into another more and more perfect up to the current human nature'.[85]

Fifteen years later in 1878, the condemnation of Darwinism was carried out this time by the Congregation of the Index itself, whose members prohibited the reading of *De' nuovi studi della filosofia*.

Discorsi a un giovane studente (*New Studies of Philosophy. Discourse to a Young Student*), by Raffaello Caverni (1837–1900), an Italian priest and professor of mathematics and philosophy at the Firenzuola Seminary in Tuscany. The censors' deliberations dealt explicitly with the fact that the book presented Darwin in a favourable manner. Their attention had been drawn to the book, whose title did not suggest in any way that it dealt with evolution, by an account published in 1877 in the journal *La Civiltà catholica*, an Italian Jesuit organ that published many texts hostile to the idea of evolution.[86] The author of the review declared in this instance that 'Darwinism is a seed of unbelief, the result of considering nature without God and the tendency to exclude God from Science. All the laws that Darwin dreamed up have the result of making divine action superfluous. Darwin scarcely concealed the equivalence of his theory and the atheistic, materialist principles that have been professed afterwards by his followers and successors without any qualification'.[87] Understandably, the book was quickly denounced in Rome. After receipt of the usual consultors' reports in favour of the censure of the book, the cardinals unanimously condemned it and showed themselves quite aware of the likely impact of their decision. The summary of their deliberations indeed indicates that, until then, the Holy See had never discussed Darwinism and that the prohibition of Caverni's book would be tantamount to doing so in an indirect manner. The decision would no doubt generate an outcry, 'the example of Galileo would be held up; it will be said that this Holy Congregation is not competent to emit judgments on physiological and ontological doctrines or theories of change'. Despite the danger of reviving the infamous 'Galileo case' that still hung over their heads like a symbolic Damocles' sword, the prelates concluded that they should not allow themselves to be influenced by this 'probable clamour', because Darwin's system 'destroys the bases of revelation and openly teaches pantheism and an abject materialism. Thus an indirect condemnation of Darwin is not only useful, but even necessary, together with that of Caverni, his defender and propagator among Italian youth'.[88] The issue was that of the autonomy of science, because the Cardinals understood that Darwin's method tends 'to exclude God from Science', an approach that could only, they believed, lead to atheism.

As for Protestants, sanctions and censorship against the proponents of evolution were conducted at a more local level, but the arguments were of the same nature as among Catholics and related, ultimately, to the question of the autonomy of science with regard to the interpretation of the Scriptures. Leaders of Protestant colleges and universities were supposed to ensure that the teaching of their professors was consistent with the religious creed of their establishments. In the event of a conflict with the directors, professors were expelled, as freedom of opinion and security of employment were non-existent in educational institutions before the middle of the twentieth century. Thus, the geologist Alexander Winchell lost his job three years after having been appointed professor in 1875 at the Methodist Vanderbilt University, in Nashville, Tennessee, when the directors of the university decided that they did not appreciate his evolutionist views.[89]

American evangelical missionaries sent to evangelize in the Ottoman Empire exported their debates on the theory of evolution into Muslim countries. In 1882, the year of Charles Darwin's death, Edwin Lewis, Presbyterian minister and professor of geology and chemistry, was shown the door by the New York board of directors of the Syrian Protestant College (the current American University of Beirut) for having presented the theory of evolution in a favourable light during the college's annual celebration that brought together students and local civic leaders.[90] A few years later, in 1884, the same scenario repeated itself in the southern United States when James Wilson, Presbyterian minister and holder of the Perkins Professorship of Natural Science in its Relations to Revealed Religion also presented a positive assessment of the theory of evolution. Summoned to explain himself to the Synod of his congregation, he insisted on the autonomy of science and, considering himself on trial, recalled Galileo's argument according to which the Bible was not intended to teach science and reiterated his belief that there was no contradiction between the evolution of species and biblical texts. But the fundamentalist current dominated and the direction of the Southern Presbyterian Theological Seminary in Columbia, South Carolina, feeling that it could not afford to lose the support of its community, dismissed Wilson in 1886.[91]

In the Catholic world, centralization in Rome made it easier to identify the many authors who promoted evolutionary theses for a general public and who were eventually censored by the Roman authorities towards the end of the nineteenth century. This was the case with Emile Ferrière's (1830–1900) work, entitled simply Le Darwinisme, published in French in 1872 and put on the Index twenty years later. Ferrière had probably attracted attention to both himself and his work with the publication of his provocative Les Erreurs scientifiques de la Bible in 1891. Indeed, the decree of 7 April 1891 that prohibited his essay on Darwin also condemned several of his previous works, including Les Apôtres. Essai d'histoire religieuse

119

d'après la méthode des sciences naturelles (1879), *L'Âme est la fonction du cerveau* (1883) and *La Matière et l'Énergie* (1887), all published by Félix Alcan. Probably in response to this condemnation, Ferrière published a follow-up to his critical work on the Bible in 1893, entitled *Les Mythes de la Bible*, that he introduced by saying:

> By decreeing that the personal God was the author of the errors of the Bible in cosmogony, Astronomy, Meteorology, Physics, Geology, Botany, Zoology and Physiology, the Council of Trent ordered one to believe on pain of damnation, that God was ignorant and an idiot. By declaring that the Bible contained the true theory of the universe in all the natural sciences, the Council of Trent doomed the Bible to ridicule and contempt.[92]

It is easy to understand that, confronted with statements such as these, the Congregation of the Index was prompt to respond and condemned the volume in the months following its publication, in a decree dated 14 July 1893, thus proving that even on the anniversary of the French Revolution, the ardour of the Roman censors was not cooled in the least.

Authors from the Catholic clergy that advanced more nuanced interpretations found just as little grace in the eyes of the Roman Curia. In 1887, the Dominican Dalmace Leroy thought it quite legitimate to freely discuss the evolution of species and published a book entitled *L'Évolution des espèces organiques*. He even predicted that the fate of this new idea would be the same as that reserved for the theses defended by Galileo: 'the idea of evolutionism will, I believe, have the same fate as that of Galileo; after having first frightened the Orthodox, once the excitement has calmed down, the truth, cleared from exaggerations on both sides, will finally emerge'.[93]

He did not say, however, if it would take years or decades ... Although prefaced by a well-known Catholic geologist, Albert de Lapparent, and by a preacher at the Notre Dame de Paris cathedral, Leroy's book did not satisfy the more conservative theologians, who believed that the text of the Bible was without ambiguity: 'the body of the first man was formed immediately from the Earth by the Creator', warned the Jesuit Joseph Brucker in his critical study of the relations between the Bible and transformism.[94] Brucker concluded his long defence of Catholic orthodoxy by saying that he hoped to have convinced scholars 'who lack theological training, that they are not fit to dogmatize on "freedom of discussion", from the point of view of faith, on question of transformism'.[95] He considered Leroy to be utterly wrong when claiming that 'Church Fathers did not want to dogmatize on the matter of evolution, on which they were not qualified'. Brucker believed, to the contrary, that science could only produce 'probable inductions' on the origins of man whereas 'full light on this matter could only come from a revelation by the Creator'. Since the 'Holy Books' effectively contained teachings relating to the origins of man, 'was it permissible', he asked rhetorically, 'for a Catholic to believe that this issue is not of the Church's competence, and that if she tries to settle it, her judgment is not infallible and binding upon all the faithful?' To say so would be to 'seriously undermine the authority the Church has received from God to determine the direction and extent of the revelations He has deigned to give us in the inspired writings'.[96] The 'conflict of the faculties' was still the order of the day at the end of the nineteenth century and the Jesuit insisted that science must remain subordinate to theology when it addressed issues that fell within the jurisdiction of the first chapters of *Genesis*.

Leroy responded to the Jesuit's attacks in 1891 in a new edition of his publication more specifically entitled *L'Évolution restreinte aux espèces organiques*, approved once again by his superiors as containing nothing that might 'offend faith or morals'.[97] Contradicting Brucker, he concluded that none of the authorities cited condemned the evolution of species and ironically added that the idea of human evolution 'is well worth the vulgar conception of a clay statue no one knows how kneaded'.[98] The polemic between the Dominican and the Jesuit continued – a clear demonstration of the internal struggles within the Church – but the Jesuit finally won, not by the probative value of his arguments but by the intervention of Roman authorities. Brucker himself admitted he did not wish to summon up the 'wrath of the Index' but the Roman Congregation responsible for the censorship of books was nonetheless called in to investigate as a result of a denunciation declaring that the book was contrary to the literal explanation of Creation contained in the Bible, still upheld by the Catholic Church. However, the first report on the book was written by a Franciscan rather favourable to Leroy, who concluded that the book contained nothing reprehensible. Dissatisfied with this report, the members of the Congregation demanded two more reports because, in reality, they wanted to prohibit the book. And even though one of the two new consultors continued to side with Leroy, the Cardinals of the Congregation decreed, in January 1895, that the book should be placed on the Index. After some discussion, it was decided that the decision would not be published if Leroy publicly retracted his views and agreed to withdraw his book from the public domain. As a good Catholic obedient to the Church, Leroy gave in. In a letter published by the Catholic daily *Le Monde* on 4 March 1895 – in a style that recalled Galileo's retraction of 1633 – he declared that he 'reject, retract, reprove all that [he] has said, written and published in favour of this thesis [that] has been examined in Rome by the competent authorities and judged untenable because it cannot be reconciled neither with the affirmations of Sacred Scripture, nor with the principles of sound philosophy'.[99] The rector of the French School of Rome, Bishop Louis Duchesne, wrote the following year to Alfred Loisy (who, as we have seen, would also suffer the wrath of the Inquisition some years later) that 'the Holy Office is not dead. One should not believe that the memory of Galileo can be useful. It protects only the chemists'.[100]

Leroy continued all the same, behind the scenes, to try to reverse the decision. A year after his retraction, the American John Zahm, member of the Congregation of Holy Cross and professor of physics at the Catholic University of Notre Dame, published *Evolution and Dogma*, translated into French the following year, professing the same ideas. Unsurprisingly, the book, also translated into Italian, was denounced to the Congregation of the Index in 1897. Aware of the balance of power within the Church between traditionalists and progressives, Zahm counted on the fact that he had recently received an honorary doctorate from Pope Leo XIII himself to bolster the legitimacy of his position in the face of conservative opposition. But that, again, overestimated the role of the Pope in the Roman bureaucracy: the Cardinals decided to place his book on the Index in September 1898. Just as they had done with Leroy, the conservatives first attempted to force Zahm to retract publicly. But since he had greater support within the heirarchy, a compromise was sought, and instead of retraction, Zahm published a letter addressed to his publishers telling them to take his book out of circulation.

As was the case for Leroy, it was then decided that the decree of the Index would not be published, so the condemnation did not become official and could not spark an embarrassing public scandal. The names 'Leroy' and 'Zahm' therefore do not appear in the official list of the Index of condemned books. This was the culture of duplicity cultivated by the Roman censorship, one that recalls the silence that had surrounded the withdrawal of Copernicus' and Galileo's books from the Index at the beginning of the 1830s. As for Zahm, he remained bitter and confided to a friend that he hoped to win this long war, convinced as he was that 'truth and justice' were on his side. It was, he pursued, 'a fight against Jesuitical tyranny, against obscurantism and medievalism'.[101] Like Leroy, Zahm would eventually be proven right, but the struggle between the conservative and progressive clans of the Catholic clergy would last another century.

At the beginning of the 1920s, the Canon Henry de Dorlodot (1855–1929), theologian and Professor of Geology and Paleontology at the Catholic University of Louvain, published a book on *Le Darwinisme d'un point de vue catholique*. He had been caught up by the open attitude of his institution that, in 1909, had delegated him as official representative to a Congress held at the University of Cambridge on the centenary of the birth of Charles Darwin. The first volume of his book focused on the origin of animal species and in it he announced a second volume dedicated to the origin of man. In the first book, Dorlodot developed ideas similar to those of Zahm and Leroy: the Catholic faith in no way prevented one from believing that the wisdom of God was able to evolve species from one or a few primitive types. As always among the Catholic supporters of evolution, this was a teleology that had nothing to do with Darwin's theory as it left in the shadows the mechanism of natural selection and the central role played by chance. But even this minimal evolutionism was unacceptable to the conservative faction of the Roman Curia and its allies.

Dorlodot's book was nonetheless lauded by the Catholic theologian Jean Rivière, a supporter of modernism who had lost his teaching position at the seminary at Albi (near Toulouse) in 1918. Rivière concluded his account of the book, published in *Revue des sciences religieuses*, by saying that he hoped that the book would contribute to eliminate the 'theses on the fixity of species that still clog some treaties *De Deo creante*'.[102] He added that he was looking forward to reading the next volume on the origin of man. But the book never appeared, because Dorlodot, like his predecessors, would be condemned to silence.

In effect, the Pontifical Biblical Commission, set up in 1902 in the context of what historians of Catholicism have called the 'modernist crisis' – which was in fact an anti-modernist reaction – had taken upon itself to evaluate the orthodoxy of *Le Darwinisme d'un point de vue catholique*. The members of the Commission asked the Rector of the University to convince Dorlodot to retract his views on evolution. Confronted with refusals from both of them to follow suit, the President of the Commission sent the file to the Holy Office.[103]

Following the usual procedure, three reports were ordered on the content of the book. In 1925, the Secretary of the Holy Office, the very conservative Cardinal Merry del Val, recounted in private to Cardinal Mercier, then prelate of Belgium, the judgment rendered by the Holy

Office. Since nothing had yet been made public, Mercier, who had been informed by the Abbot Henri Breuil (1877–1961), a prominent prehistorian, of the danger that such a condemnation would cause to the reputation of Catholic scholars, lobbied the Roman Curia to abandon the procedures. The rumour had circulated for some time that Rome was preparing to condemn all theories of evolution in a sort of new *Syllabus*.[104] It eventually proved unfounded and, in November 1925, the Cardinal, reassured by the Roman authorities, wrote to the Rector of the Catholic University of Louvain, an institution in which he had long been a professor of philosophy, to say that 'the silence of the last months has been a good sign. Let us hope it lasts'.[105] In the end, the book was never condemned and Dorlodot did not retract. Faithful to his strategy of 'combining prudence and firmness', as he revealed to a friend in October 1925, he decided, however, not to publish his proposed work on human evolution.[106] He considered that 'the moment had not arrived' and that his decision to remain silent was bolstered 'by the unimaginable condemnation by the Index of Brassac's *Manuel biblique*'.[107]

Abbot Auguste Brassac had contributed to the new edition of the *Manuel biblique* produced by the Sulpicians Louis Bacuez (1820–1892) and Fulcran Vigouroux (1837–1915), a book reissued many times since the end of the nineteenth century. Yet its updated edition had been considered much too favourable to critical history, which, as we have seen above, had been condemned by the Church in 1907. For intransigent Catholics – whom the more liberal did not hesitate to describe as 'fundamentalists' – the condemnation of the new edition of the manual in 1924 reaffirmed the 'necessary subordination of exegesis (and of history) to theology'.[108] In the summer of 1924, Brassac confided to a friend that Pius XI, elected in 1922, had, for a long time, been tired and that several cardinals including Merry Del Val and 'the extreme right [have] take[n] this opportunity to impose their ideas'.[109]

Dorlodot was more orthodox than modernist and proposed a concordist interpretation based on a reading of the Church Fathers. A review of the English translation by a member of the Congregation of Holy Cross also concluded that the book offered nothing new in comparison to that of his coreligionist, Father John Zahm, published more than twenty years earlier and that we discussed above.[110] But this new attempt to harmonize the interpretation of the Bible and the evolution of animal species was still too heretical for the cardinals of the Biblical Commission and the Holy Office. Thus, in June 1923, the rather conservative Cardinal Van Rossum wrote to the rector of the Catholic University of Louvain to remind him of 'the various measures that the Holy See has taken in the last thirty years to stop the spread of Darwinist theories among Catholics'.[111]

In such a context, the pause in the proceedings against Dorlodot suggests a respite in the conflict between the conservative faction and the Catholic Darwinians, as Pius XI was probably occupied with more urgent tasks – such as the recognition of the Vatican State in 1929 – than with debates surrounding the theory of evolution. In the years that followed, the Catholic proponents of evolution no longer seem to have been hounded by the Congregations of the Holy Office and Index and discussions of human evolution were tolerated. As such, Ernest Messenger, a priest who had studied theology at Louvain and had translated Dorlodot's book into English early on in 1922, developed the latter's ideas further in a 1932 essay entitled *Evolution and Theology: The Problem of Man's Origins*, without incurring the wrath of the religious authorities.[112]

That being said, the Congregation of the Index continued its work and did not abstain from condemning the evolutionary works of the spiritualist philosopher Édouard Le Roy. His essays on *L'Exigence idéaliste et le fait de l'évolution* (1927) and *Les Origines humaines et l'évolution de l'intelligence* (1928) were placed on the Index in 1931, as was *Creative Evolution* by his friend and intellectual guide, the philosopher and Nobel Prize Laureate Henri Bergson, that had been condemned in 1914.[113] Another friend, the Jesuit Teilhard de Chardin, a supporter of a theist evolutionism, who had been removed from his chair as Professor of Geology at the Catholic Institute of Paris in 1925 for his avant-garde ideas in theological matters, had helped Le Roy in the drafting of the second book. Upon hearing of the condemnation, he wrote to Le Roy from his exile in China saying that he too felt slighted and that this decree, issued by 'the narrow powers of the Roman Orthodoxy' betrayed such an enormous ignorance of Le Roy's thought that he was 'truly bewildered'.[114] Le Roy, who succeeded Bergson in the Chair of Modern Philosophy at the Collège de France in 1921, was part of a network of Catholic evolutionist scientists. He would play a central role in 1929 in the creation of a chair in prehistory for Abbot Henri Breuil at the College.[115]

A world-renowned scientist, Henri Breuil (1877–1961) was nevertheless viewed with suspicion by religious authorities. When Pius XI announced the foundation of the Pontifical Academy of Sciences, in 1936, Breuil, like Teilhard de Chardin, was on the list of potential members. However, the investigative report on Breuil's orthodoxy noted that the issues with which he dealt 'are among the most delicate, as these touch on the origins of humanity and races'. Although he remained close to the facts and did not indulge in 'general theories', he was known as a supporter of 'the hypothesis of preadamite humans'. A prudent man, Breuil had abstained from publically discussing the possible consequences of the evolution of man on the question of original sin, in contrast to the more adventurous Teilhard de Chardin. The report concluded that it fell to the Holy Father 'to judge if one does not run the risk, in naming him, of appearing to give a pass to a doctrine whose theological consequences are serious'.[116] Like Teilhard, Breuil was struck off the list of the first members of the Pontifical Academy, whose creation was meant to signal an opening to modern science and which, in its inauguration in 1937, received from the Pope the mission to . . . 'serve the truth'.[117]

Following this anti-modernist period, the mentality of the Roman authorities gradually evolved. In 1950, Pius XII finally conceded in the encyclical *Humani generis* 'Concerning some false opinions threatening to undermine the foundations of Catholic doctrine' that 'the Church does not forbid that, in conformity with the present state of human sciences and sacred theology, research and discussions, on the part of men experienced in both fields, take place with regard to the doctrine of evolution, in as far as it inquires into the origin of the human body as coming from pre-existent and living matter'. Some commentators like to believe that the Church was, from then on, no longer opposed to the theory of evolution, but they tend to forget to quote the rest of the text: '*provided that all are prepared to submit to the judgment of the Church*, to whom Christ has given the mission of interpreting authentically the Sacred Scriptures and of defending the dogmas of faith'.[118] Moreover, the Church still considered that '*both opinions*, that is, those favourable *and those unfavourable to evolution*' were still on the same footing in 1950, whereas the fixity of species had then been eliminated from science for more than half a century. Finally, it should not be thought that the door was open to *all* scientific hypotheses, for the Encyclical had explicitly noted that the polygenist hypothesis, according to which humans might have descended from several different lines, was not acceptable and that on this question 'the children of the Church *by no means enjoy such liberty*'. Polygenism was indeed inconsistent with the teachings on the original sin, which directly connects all humans to Adam, the first man. Thus the 'faithful cannot embrace that opinion which maintains that either after Adam there existed on this Earth true men who did not take their origin through natural generation from him as from the first parent of all, or that Adam represents a certain number of first parents'. For the Catholic Church, the original sin was 'actually committed by an individual Adam and which, through generation, is passed on to all and is in everyone as his own'. Ten years after this publication, a Cardinal reminded Catholic scientists that it would therefore be 'reckless for a Catholic to deviate from that judgment and the scientist must take it into account, not as a bias that hinders the objectivity of his research, but as something superior that, although foreign to science, can help him subjectively'.[119]

The theory of evolution continued to make inroads within the Catholic Church whose changes in thought moved at glacial speed and, in 1996, John Paul II finally declared that, 'more than a half-century after the appearance of that encyclical [*Humani generis*], some new findings lead us toward the recognition of evolution as *more than an hypothesis*'.[120] More than a hypothesis: in the cryptic language of the Vatican scribes, this was as far as the Church could go in the admission of an error that had lasted for a more than a century.

The end of the twentieth century thus witnessed a reversal in the balance of power between Christian theology and science. In his speech before the Pontifical Academy of Sciences in 1992, John Paul II stated that it was 'a duty for theologians to keep themselves regularly informed of scientific advances in order to examine, if such be necessary, whether or not there are reasons for taking them into account in their reflection or for introducing changes in their teaching'.[121] Four years later, before the same Assembly, he repeated that 'theologians and those working on the exegesis of the Scriptures need to be well informed regarding the results of the latest scientific research' in order to '*set proper limits to the understanding of Scripture*, excluding any unseasonable interpretations which would make it mean something which it is not intended to mean'.[122]

The Church thus finally conceded that Galileo was right when, in 1615, he said, as quoted above, that 'the intention of the Holy Spirit is to teach us how one goes to heaven and not how heaven goes'.[123]

— 5 —

FROM CONFLICT TO DIALOGUE?

In reality, between religions and real science there exists neither relationship nor friendship, nor even enmity; they live on different planets.

Friedrich Nietzsche[1]

According to most historians of science of the last twenty years, the idea of a 'conflict' between science and religion would have arisen only in the last quarter of the nineteenth century with the publication of the works of the Americans John William Draper (1811–1882), doctor and chemist, and Andrew Dickson White (1832–1918), historian and first president of Cornell University. In 1874, Draper published his *History of the Conflict Between Science and Religion*; White's book, entitled *A History of the Warfare of Science with Theology in Christendom*, appeared in 1896. According to Ronald Numbers, who provided a striking summary of this now widespread thesis, these two authors are at the origin of 'the greatest myth in the history of science and religion', one that 'holds that they have been in a state of *constant* conflict'.[2]

In presenting these works as constitutive of the 'myth' of a conflict between science and religion, the many historians of science that repeat this story thus suggest that there was no real conflict between science and religion and that statements by the founders of the 'myth' were not only often false, but always exaggerated. These same historians furthermore claim that the conflicts were neither constant nor inevitable. First, it seems at best dubious that conflict was not inevitable, given that the previous chapters have amply demonstrated that disciplines such as cosmology, geology and biology did raise questions relating directly to the biblical texts that religious fundamentalists interpreted (and continue to interpret) literally. They thus could not ignore contrary interpretations, given the superiority they endowed religion over science. The collisions were in fact predictable and have even been predicted by many of the actors involved in the different conflicts that opposed science and religion. Second, that the conflicts were not 'constant' is obvious as nothing in history is 'constant': people die and others lose interest in old debates. And while Numbers admits that '*discussions* of the relationship between "science" and "religion" originated in the early nineteenth century' and that there were 'occasional *expressions of concern* about the *tension* between faith and reason', he also claims that '*no one* pitted religion against science or vice versa'.[3]

This surprising interpretation has been regularly reiterated by historians of science over the last twenty years. But using euphemisms like 'discussion', 'tension' and 'expressions of concern' to try to make even the most obvious conflicts disappear, just denies the evidence and does not stand up to critical examination. Negating the existence of conflict, however, is consistent with an idea dear to this historiography that consistently seeks to minimize the most flagrant cases of conflict, *conflicts that were perceived as such by the actors of the time*, by reducing them to the rank of exceptions. A psychological reductionism leads Numbers to suggest, for example, that Draper's book can even be explained by his personal conflict with his sister who converted to Catholicism.[4] Yet it seems rather more plausible, following Glenn Altschuler, to situate this book in the broader context of a nineteenth century marked by a questioning of Christianity caused by the development of science and critical history. Above all, Numbers' psychological reductionism does not explain the fact that Draper's book was immediately translated into several languages, including French, in the year following its release in 1874, a fairly clear sign that the book had a social relevance that was much broader than a simple conflict with his sibling.[5] Finally, it is curious that a number of authors who enjoy repeating the story of Draper and White, studiously avoid mentioning the fact that Draper's book was put on the Index by the Catholic Church in 1876.[6]

In light of the central role given to these two authors in the origin of the 'conflict thesis' (as some historians call it), let us return briefly to the context in which their books were written. As regards White, his book was a response to the Protestant communities opposed to the non-denominational character of the newly founded Cornell University that opened in 1868 under White's presidency, and that he would lead for twenty years.[7] White explained in the introduction to his book that after having adopted a strategy of conciliation that had not borne fruit, as neither side wished to compromise, he decided to go on the attack in a speech given in December 1869. Entitled 'The Warfare of Science', it was immediately published in a national daily newspaper. Using the same title, White developed his thesis in two articles a few years later, in February and March 1876, in *Popular Science Monthly*.[8] These papers were then immediately collated into a book. Two years earlier, the same magazine also published the preface to Draper's book. In 1890, White returned to the question in a long series of articles, published in installments that appeared up until 1895 in the same magazine under the title 'New Chapters in the Warfare of Science', in which he went through a number of disciplines (archeology, anthropology, medicine, astronomy, etc.), discussing their relations with Christianity. Together, these articles led eventually to the publication of his massive two-volume book in 1896. Like Draper's, White's book was rapidly translated into several languages (German, Italian, Swedish and French), a sign that the topic had touched a sensitive nerve in many countries.[9]

In addition to restricting themselves to individuals and their psychology and thus ignoring institutions, the work of (mostly Anglo-American) historians of science and religion over the last twenty years has fallen prey to the important methodological weakness of confusing the points of view of the analyst and those of the actors as if they were on the same plane. In doing so, they set themselves up as spokespersons or advocates for the positions maintained by some actors (those who have an interest denying or minimizing conflict) and not others, instead of simply observing the debates without taking sides, to understand why it was going on. As a consequence, they often merely restate the point of view of historical actors instead of analysing them. For example, the editor of *Popular Science Monthly* had already noted, in the months that followed the publication of Draper's work in 1874, that the representatives of various religious denominations (Jewish, Unitarian and Catholic) all argued that such conflicts did not exist and indeed had never existed.[10] Similarly, the Canadian geologist John William Dawson intervened in the debate in 1876 claiming that the conflict between science and religion was little more than the result of a misuse of words because 'true religion, which consists in practical love to God and to our fellow-men, can have no conflict with true science'. Dawson even underlined an aspect often ignored by our historians – who limit themselves to an individualistic conception of religion – when he recalled that '*the actual conflict of science*, when historically analysed', was not with religion per se but was rather 'fourfold: 1. With the Church; 2. With theology; 3. With superstition; 4. With false or imperfect science and philosophy'. In sum, even the very pious Dawson (a Protestant who viewed the Catholic religion with horror) clearly distinguished between *individual* beliefs and *institutions* (the Church and its theologians).[11]

This amalgamation of levels of analysis also leads historians to speak of 'models' of conflict, harmony or independence between science and religion, which might then be imposed on the historical 'data'. These models, however, are never clearly distinguished from the points of view of the historical actors who have their own conception of the relationships between science, religion and theology, also founded on conflict, harmony or independence.[12] However, it is not a question of choosing between these models, as some historians seem to think when they promote a vague form of ecumenism devoid of any historiographical interest, but to analyse what the actors say of this relationship. For example, it is easy to disqualify the historical narratives proposed by Draper and White, as John Brooke, Ronald Numbers and many others have done, since it is obvious that for nineteenth-century writers like Draper and White, history was not an academic discipline but rather a weapon in their ideological struggle against a religious power that *they* considered obscurantist.[13] There was no place for nuance in a debate that, like all debates, favoured polarization. In fact, Draper himself said so explicitly in the preface to his book:

> In the management of each chapter I have usually set forth the ortho-dox view first, and then followed it with that of its opponents. In thus treating the subject it has not been necessary to pay much regard to the moderate or intermediate opinions, for, though they may be intrinsically of great value, in conflicts of this kind it is not with the moderates but with the extremists that the impartial reader is mainly concerned. Their movements determine the issue.[14]

A final element in the tendency to deny the existence of open conflict between science and religion throughout the nineteenth century is a kind of methodological blindness to the reality that when social actors insist that there is no conflict on a given question, this means in fact that there is at least a perception of conflict in the surrounding milieu and that they intend to change that perception. The simple fact that individuals, in different historical periods, say that there is no conflict between science and religion clearly indicates that they assume that other people believe that such a conflict exists. If it had been self-evident to all that science and religion were either independent or inseparable, then there would simply have been no discourse or publications *promoting* 'harmony' between the two.

The fact that there is a perception of conflict among many actors since at least the beginning of the nineteenth century is therefore indisputable, as the previous chapters have amply shown. We may thus legitimately speak of a history of conflict between science and religion, without siding with one of the factions or – worse – trying to go 'beyond war and peace', as suggested by Lindberg and Numbers, the two most active historians in this area.[15] As the historian Marc Bloch said, when the historian has observed and explained, his task is finished. Only the judge has to make a judgment and deliver a sentence.[16]

In sum, and contrary to the now dominant trend in the historiography of the last twenty years, the study of the history of conflict between science and religion should not seek to partake in the debates that oppose the different factions in this disputed terrain, but should, more simply, follow the discussions and the actors in order to see who speaks of conflict and in what context. As we will see in greater detail in this chapter, the idea of conflict, potential or real, between science and religion is clearly present in the literature since at least the 1820s. Draper and White came, in fact, somewhat late to the game. They were more radical and polemical in affirming the existence of a recurring debate but they did not invent the theme. Those who criticize the notion of 'conflict' have often suggested that it is the fruit of an 'essentialist' conception of science and religion that views the two as unchanged for 400 years. These critics cite no precise examples for the obvious reason that, of course, nobody has ever explicitly supported such a simple-minded thesis that 'science' did not change between the seventeenth and the nineteenth centuries.[17] There is no need to be 'essentialist' – a term always used more as an insult than a philosophical concept – in order to identify, in a given time frame, the social roles of actors and institutions. It is obvious that the practice of science at the beginning or even at the end of the seventeenth century differs from that in the middle of the nineteenth century. But Galileo and Kepler knew very well that they were doing astronomy and not theology, and they had no difficulty in identifying the theologians with whom they debated the autonomy of their science, as we have seen in the previous chapters.

Given these methodological considerations, let us examine the content of the discourse on the question of the relations between science and religion to see whether the actors did talk of 'conflict' or not. We will return later to the reasons that have contributed to the development of the now dominant historiography that we can qualify as ecumenical, insofar as it tends to minimize the existence of conflict and to insist on the notions of 'meeting', 'encounter' and 'dialogue'.

The evolution of the discourse on the relationship between science and religion

Let us begin with a measure: the evolution of the use of the words 'science' and 'religion' in the anglophone and francophone public spaces since the beginning of the nineteenth century. One way to embrace the whole of public discourse is to analyse the content of the books and articles published in the nineteenth century. Using the database of the millions of books digitized by Google Books, we can generate a roughly representative sample (though certainly not exhaustive) of the state of public discourse over the years.[18] As figure 5.1 shows, in the anglophone world (essentially Great Britain and the United States), the term 'religion' was, until the 1930s, much more frequent, in proportion, in all publications, than the term 'science'. Subsequently, 'science' takes over, but, beginning in the 1980s the use of 'religion' rose again while that of 'science' stagnated, so that the two curves begin to converge at the end of the 1990s.

Interestingly, the relative importance of the two words in public discourse is very different in the French language corpus. As figure 5.2 shows, the influence of rationalism after the French Revolution and the rise of positivist philosophy in the middle of the nineteenth century clearly had an effect on usage. It is striking that the reversal of the relative importance of the terms 'science' and 'religion' took place towards 1850 in the francophone world, that is three-quarters of a century earlier than in the anglophone world within which the many Protestant sects remained culturally important. Despite this difference, the return of religious discourse is also visible on this graph, since there is a clear escalation of the theme of religion beginning in the 1980s, the two curves joining again toward the end of the 2000s.

Let us now look at the evolution of public discussion on the relationship between science and religion using once again the corpus of works digitized by Google Books. What strikes one first in the anglophone corpus (figure 5.3), is the steady rise of the expression 'science and religion' from 1800 to 1880, followed by a stabilization between 1880 and 1920, and a sudden but temporary resumption between 1920 and 1940. After the Second World War, there is a general decline, followed by continuous growth since the 1980s.

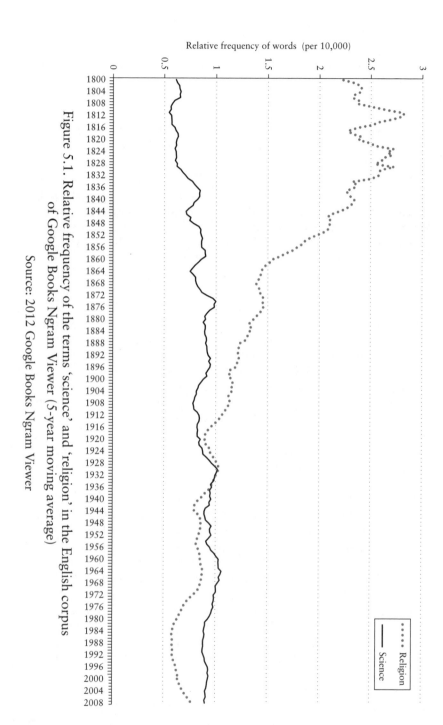

Relative frequency of words (per 10,000)

Figure 5.1. Relative frequency of the terms 'science' and 'religion' in the English corpus of Google Books Ngram Viewer (5-year moving average)

Source: 2012 Google Books Ngram Viewer

•••••• Religion

———— Science

134

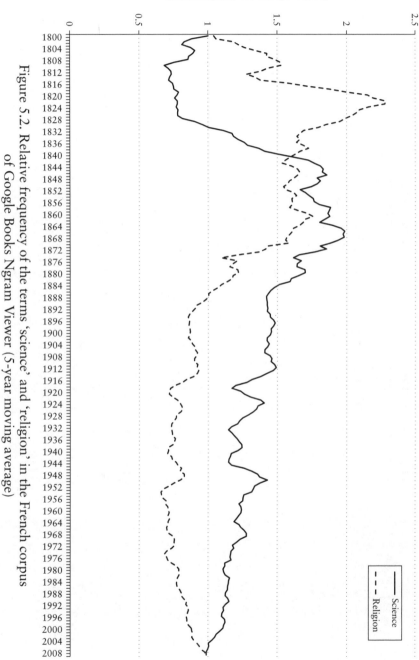

Figure 5.2. Relative frequency of the terms 'science' and 'religion' in the French corpus of Google Books Ngram Viewer (5-year moving average)

Source: 2012 Google Books Ngram Viewer

135

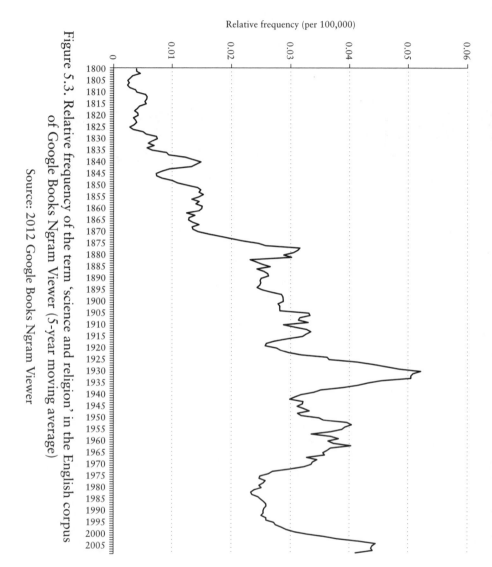

Figure 5.3. Relative frequency of the term 'science and religion' in the English corpus of Google Books Ngram Viewer (5-year moving average)

Source: 2012 Google Books Ngram Viewer

136

On a smaller scale, we note growth from the beginning of the 1830s, a period that coincides with the creation of the British Association for the Advancement of Science (BAAS) in 1831. The promotion of science in England then increased in social visibility, which in turn could not help but raise the question of its relationship with the then dominant religious discourse. The second growth spurt began at the outset of the 1870s, obviously with the publication of Draper and White's critical essays. The third upsurge corresponds to debates arising between the two wars involving authors such as Bertrand Russell who published *Religion and Science* in 1935.[19]

On the francophone side (figure 5.4), the situation is once again slightly different. Interest in the relationship between science and religion gained importance only at the beginning of the 1860s, in the context of Darwinism and as part of the controversy over Renan's *The Life of Jesus*.[20] Subsequently, there was a steady rise, with plateaus in the years 1920–1940 and 1960–1980, followed by a resurgence since the 1980s. It should be noted that the collective organization of French science was established later than in England; the French equivalent of the BAAS, the French Association for the Advancement of Science (AFAS), was not founded until 1872.[21] In addition, natural theology was less popular in France than in the anglophone world, which clearly affected public discourse. Thus, for example, the large survey of members of the Academy of Sciences on 'religious feeling and science', published in *Le Figaro* throughout the month of May 1926, does not seem to have generated debates as rich as those that took place in England in the early 1930s.

The evolution of the number of books containing the terms 'science' and 'religion' in their title (figure 5.5) provides another interesting indicator confirming the escalation of interest in the question of the relationship between science and religion.[22] All things being equal, the fact that most of the books that have flooded the market, particularly since the 1990s, are in English confirms the cultural difference that separates the anglophone and francophone worlds on this question.

In general, we can say that societal interest in the question of the relationship between science and religion first arose during the first half of the nineteenth century. The sudden upsurges of interest that followed correspond to periods of intense debates that are explained by local events that were situational and often contingent, as was the case in France in 1926 and in England at the beginning of the 1930s.

We now come to the idea of 'conflict between science and religion', considered as a 'myth' by most of the historians who, since the

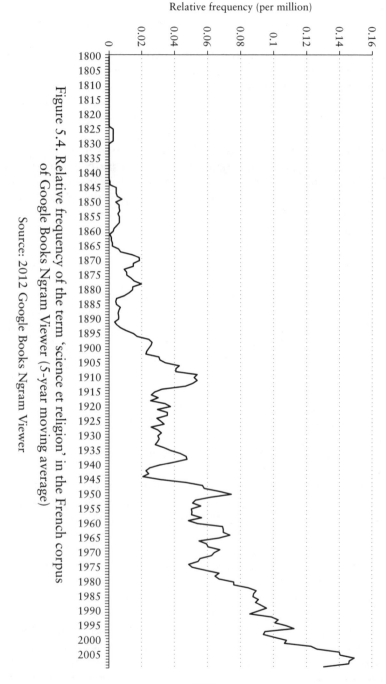

Figure 5.4. Relative frequency of the term 'science et religion' in the French corpus of Google Books Ngram Viewer (5-year moving average)

Source: 2012 Google Books Ngram Viewer

138

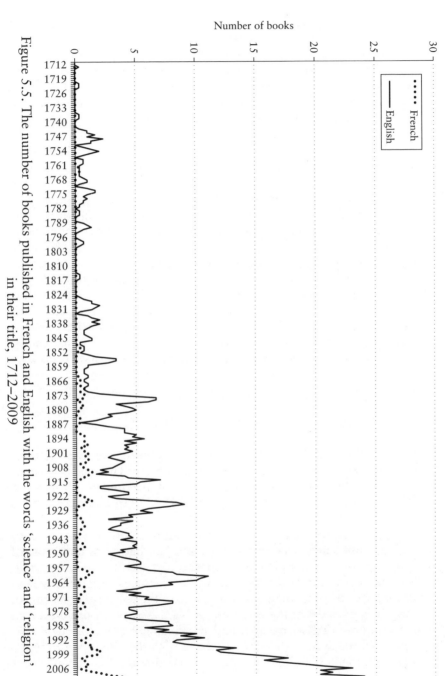

Figure 5.5. The number of books published in French and English with the words 'science' and 'religion' in their title, 1712–2009

Source: Catalog of the Widener Library, Harvard University

1980s, have written on the relationship between these two social fields. While recent historians talk of 'encounter', 'meeting' and 'conversation', we are here interested in the very word 'conflict' as used by social actors since the beginning of the nineteenth century. We will thus be able to see whether or not anybody did indeed speak of 'conflict' and not only of 'tension' or 'concern'. As figure 5.6 shows, there was indeed an upsurge of interest in debating the question of 'conflict between science and religion' in the corpus of English books at the beginning of the 1870s, in obvious response to White's first polemic intervention in 1869, and Draper's piece in *Popular Science Monthly* in 1874, as well as his book published the same year. But the notion of 'conflict' itself did not originate there but much earlier as we will see later on. The debate remained relatively stable during the period 1880–1920. It seems much more stormy in the anglophone world than in the francophone world that indeed saw the discourse on 'the conflict between science and religion' increase in frequency after 1945 only to quickly subside at the beginning of the 1960s. It did not resurface as a hot topic until the end of the 1980s, as one would expect in light of the data in the previous figures.[23] The vertical scale of the figure does not allow one to see that the idea of conflict emerged in fact during the 1820s and 1830s before experiencing a rapid rise during the 1870s. After having counted these books, let us now look at their content.

The rise of conflict

One of the first books to address the question of the relationship between science and religion was that of Thomas Dick entitled *The Christian Philosopher, or the Connection of Science and Philosophy with Religion*, published in 1823.[24] Interestingly, Dick's book does contain a section entitled 'Bad Effect of Setting Religion in Opposition to Science', in which the author criticizes 'some theologians' (unnamed) who demean the study of nature as unworthy of contemplation. According to such theologians, 'to be a bad philosopher [of nature] is the surest way to become a good Christian'. Listening to these theologians would lead to the conclusion, he added, that 'they considered the economy of Nature as set *in opposition to* the economy of Redemption, and that it is not the same God that contrived the system of Nature, who is also the "Author of eternal salvation to all them that obey him"'.[25] Further on, Dick detailed this opposition:

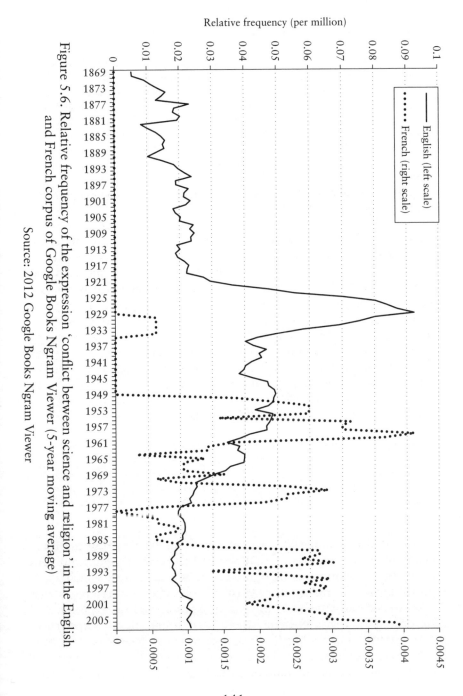

Figure 5.6. Relative frequency of the expression 'conflict between science and religion' in the English and French corpus of Google Books Ngram Viewer (5-year moving average)

Source: 2012 Google Books Ngram Viewer

141

The Philosopher [of nature] has occasionally been disposed to investigate the economy of nature, without a reference to the attributes of that Almighty Being who presides over its movements, as if the universe were a self-moving and independent machine; and has, not infrequently taken occasion, from certain obscure and insulated facts, to throw out insinuations *hostile to* the truth and the character of the Christian Revelation. The Theologian, on the other hand, in the heat of his intemperate zeal against the infidel philosopher, has, unguardedly, been led to *declaim against* the study of science, as if it were *unfriendly to religion* – has in effect, set the works of God *in opposition to* his word – has confounded the foolish theories of speculative minds with the rational study of the works of Deity – and has thus prevented the mass of mankind from expanding their minds, by the contemplation of the beauties and sublimities of nature.[26]

Dick then declared that it was 'now high time that a complete *reconciliation* were effected between these contending parties.' The expression chosen by the author ('*contending parties*') clearly indicates that it was very much a question of conflict and struggle between two options perceived by many as opposed.

A few years later, reacting to the activities of the newly created BAAS, Frederick Nolan, Vicar of Prittlewell, in Essex, in a series of lectures given at Oxford University in 1833 denounced the danger that this propaganda represented for religion and amply discussed the 'opposition' (his term) between these two areas:

If we, therefore, intend, that religion should continue to maintain its sway, some effort must be made to conciliate those interests, *which cannot remain opposed*, without endangering its very existence. Where Science is observed to assume an aspect *hostile* to Revelation, if we cannot strengthen ourselves by its alliance, we must at least endeavour to *render it neutral*. If this object be not attainable, it can be scarcely matter of doubt, that as the claims of Science advance, the interests of morality will decline, with the decay of Religion.[27]

His first lecture opened, moreover, with an extract from the Bible (Timothy, 6: 20–21) stating that one had to avoid 'profane and vain babblings, and oppositions of science falsely so called: which some professing have erred from the faith', which suggests a certain distrust of a science that seemed to draw its practitioners away from faith.[28] According to Nolan, in effect, it had been 'found, by sad experience, that from no quarter has Revelation suffered so deeply in its credit, as from *opposing* Science'. He clearly understood that from the point of view of science, 'all that bears the air of the marvellous, in its

142

deviation from natural truth, is necessarily *opposed* to science; and, when estimated by this standard, incurs the imputation of error'. He immediately added, however, that 'as the inspired record derives chiefly, if not exclusively, from the marvellous, the proof of its divine original; when that character is impeached or forfeited, the authority is undermined, which qualifies it to become the undeviating rule of our opinions and practice'.[29] In fact, Nolan rejected as contrary to a healthy philosophy the elevation of secondary causes (natural), over first causes (supernatural). Without the latter, science remained, he believed, empty and without foundation.[30] When science is pursued on its own, he added, it becomes pointless to deny that religion has been tacitly abandoned.[31] This idea that science can lead to the negation of religion and god is very old and can be found in Plato who wrote in *The Laws* that ordinary people believe that those who study the world with 'the help of astronomy, and the accompanying arts of demonstration, may become godless, because they see, as far as they can see, things happening by necessity, and not by an intelligent will accomplishing good'.[32] Of course Plato, the first of the natural theologians, opposed the 'ordinary' view; for him, astronomy helped discover nature as ruled by Mind (god) and not by 'necessity', and this is why he wanted rulers of the city to study astronomy.

The response to Nolan's attack targeting the young BAAS came quickly in a book published in 1833, *Revelation and Science*, by Baden Powell, Professor of Geometry at Oxford and a member of a network of scholars who promoted the sciences within the BAAS. As a sign that the idea of conflict was then in the air, the *British Magazine* published a brief account of the work; the author (anonymous) perceiving in this controversy 'every symptom of the commencement of a pretty hot warfare'. Confessing being himself too prudent to add anything on the subject, he concluded that one should leave to 'Mr. Powell and Dr. Nolan to fight the battle out'.[33] Powell's position was simple: science cannot be limited by anything drawn from the Bible.[34]

In another example of what we can call the 'conflict discourse' on science and religion, dating from the end of the 1830s, the author of an anti-Newtonian book quoted (without giving the exact reference) an article from the *Edinburgh Review* commenting on recent geological discoveries and that claimed that:

> The few who were first admitted to the secrets [of geology], anticipated *the conflict between science and religion*; and dreaded that the geologist, like the astronomer, might be summoned to the bar of some modern inquisition. Conscious, however, that one truth could never be

at variance with another, the patient geologists pursued their labour; and in less than half a century they have created a new department of knowledge, which, in point of philosophical and scientific unrest, will not yield to the most exalted of the physical sciences.[35]

These authors were not, of course, the only ones to perceive the dangers of natural science for religion and to talk of 'conflict'. We saw in chapter 3 that, at the same time, the Reverend John Henry Newman (not yet converted to Catholicism) had attracted attention as a result of his remarks on 'The Usurpation of Reason' (1831) and on 'Faith and Reason, Contrasted as Habits of Mind' (1839). In 1837, the North American Review published a review of a book by Hubbard Winslow titled 'The Relation of Natural Science to Revealed Religion'. The anonymous summary suggested that the book, which was initially a talk presented to the members of the Natural History Society of Boston, sought to 'to *show the harmony* between the facts of the natural world and the doctrine which Christians believed to be revealed from heaven' and that 'science and religion *are not enemies* but inseparable friends'. In reviewing the various sciences (cosmogony, geology, natural history, natural philosophy, etc.), Winslow suggested that the facts they had brought to light accorded with the teachings of the Bible. The author of the North American Review welcomed this idea and concluded that with time 'our citizens will be so well convinced of the alliance between the study of natural history, and the cause of religion' that they would generously support the scholarly societies dedicated to the progress of science.[36] This example shows that the discourse insisting on the compatibility of religion and science sought to ensure the development of science in society. Unchanged, discourses like these would resurface in the 2000s when – as we will see later in this chapter – organizations like the American Association for the Advancement of Science (AAAS) would also promote a 'dialogue' between science and religion to try to contain and calm down religious criticisms of science.

So as not to uselessly (and boringly) multiply examples that clearly show that the idea of conflict is not a 'myth' invented by Draper and White but truly a feeling shared by many social actors (scientific and religious) between 1820 and 1860, thus well before the intense conflict caused by their two polemical books, we will limit ourselves to a final quote from a lecture that Cardinal Newman (now turned Catholic) delivered in November 1855 to the students of the School of Medicine on the occasion of the opening of that Faculty in the still very young Catholic University of Dublin. After some introductory remarks on the choice of his lecture theme 'Christianity and physical science', he presented his thesis:

> I propose, then, to discuss *the antagonism which is popularly supposed to exist between Physics and Theology*; and to show, first, that such antagonism does not really exist, and, next, to account for the circumstance that so groundless an imagination should have got abroad.
>
> I think I am not mistaken in the fact that *there exists, both in the educated and half-educated portions of the community, something of a surmise or misgiving, that there really is at bottom a certain contrariety between the declarations of religion and the results of physical inquiry*; a suspicion such, that, while it encourages those persons who are not over-religious to anticipate a coming day, when at length the difference will *break out into open conflict*, to the disadvantage of Revelation, it leads religious minds, on the other hand, who have not had the opportunity of considering accurately the state of the case, to be jealous of the researches, and prejudiced against the discoveries, of Science. The consequence is, on the one side, a certain contempt of Theology; on the other, a disposition to undervalue, to deny, to ridicule, to discourage, and almost to denounce, the labours of the physiological, astronomical, or geological investigator.[37]

A generation before Draper published his pamphlet, Cardinal Newman had clearly perceived that the idea of a conflict between science and religion was well anchored in both the minds of the educated and those who were less educated and he clearly articulated the reasons behind that situation. Of course, he held that these conflicts could be avoided if each remained within the specific limits of its domain:

> If, then, Theology be the philosophy of the supernatural world, and Science the philosophy of the natural, Theology and Science, whether in their respective ideas, or again in their own actual fields, on the whole, are incommunicable, incapable of collision, and needing, at most to be connected, never to be reconciled.[38]

Newman was opposed not only to natural theology, as we saw in chapter 3, but also to the concordist stance that sought to reconcile science, theology and religion. But what interests us here is to observe that the idea of conflict was clearly articulated, and this, well before the invention of the alleged 'myth' or 'model' of conflict by Draper and White in the last quarter of the nineteenth century. Unlike historians who do not distinguish between individual and institutional points of view, Newman clearly understood this difference and even asserted that, when a scientist is religious, his opinion of God is 'private, not

145

professional, the view, not of a physicist, but of a religious man; and this, not because physical science says anything different, but simply because it says nothing at all on the subject, nor can do so by the very undertaking with which it set out'.[39] Physicists who have published books on 'God and the new physics' since the 1990s and who attempt to reburnish an old-fashioned natural theology might be interested in carefully reading the venerable Cardinal Newman.

The postulate of the separation of domains suffered several exceptions, however, and Newman knew very well that the Bible also contained statements that appeared to fall within the factual domain of physics. He believed, nonetheless, that he could easily solve the problem by noting that the Catholic Church had not, on these points, ever set out an official interpretation to which believers would have to commit. In such a context, it would be presumptuous to say what such or such a statement of the Bible really meant. He therefore considered it unlikely that a discovery in physics could be incompatible with all the possible meanings that one could assign to a biblical statement. Having admitted the existence of a possible common ground where the two areas could 'fight a battle', Newman therefore concluded that even on this point there was no reason to expect a real conflict.[40]

As he himself noted, after having shown that physics and theology could not enter into conflict, his presentation should have stopped there. However, one problem remained: how could it be explained that, despite this theoretical impossibility, the fact was that theologians and physicists did continue to squabble? According to Newman, these conflicts arose from a desire to apply the methods of theology to science or those of the sciences to theology, thus illegitimately trespassing the boundaries of each discipline.[41]

Twenty years after Newman's presentation, Draper and White's radical interventions gave the debate much greater public visibility, as evidenced by the rapid rise in the number of works dealing with the conflict between science and religion during the 1870s. The scientific context was very different from that of the 1850s, as Charles Darwin had in the interim published *On the Origin of Species* (1859) and triggered an ideological storm. The religious context had also changed, in part because of the rise of Darwinism. As we saw in the previous chapter, Pius IX published his *Syllabus of Errors* in 1864 and the Council of Vatican I confirmed, in 1870, the infallibility of the Pope and the inerrancy of the Bible – everything to enrage rationalist minds calling for total freedom of research on questions falling within the scope of science.

The two books of God 'cannot contradict each other'

Scientists, as well as educated men and women of the nineteenth century, were perfectly aware of the existence of real and potential conflicts between some of the sciences, such as geology and natural history, and some biblical statements. But this in no way prevented them from believing that such conflicts could be avoided and were, in a sense, false conflicts. The vast majority were sincere Christians, convinced that there could be no real contradiction between the two types of truth. From a rhetorical point of view, all the arguments put forward to counter the idea of real conflict between science and religion can be reduced to two basic types, based on the following assumption: because God is at the origin of the two 'books' – the Holy Scriptures and Nature – there could be no contradiction between them, since God (being perfect) could obviously not contradict himself. This idea goes back to the beginnings of Christianity and was resuscitated whenever there was a debate on the apparent disagreement between two readings, as we have already seen on several occasions in previous chapters. In his encyclical *Fides et Ratio*, (Faith and Reason) of 1998, John Paul II, quoting Thomas Aquinas, again reaffirmed that 'both the light of reason and the light of faith come from God [. . .] hence there can be no contradiction between them'.[42]

The first type of argument invoked to deny the existence of any real conflict is that such apparent conflict is only the effect of a 'false' science; when the latter cedes its place to 'true' science, the conflict with religion, considered, from this point of view, as providing the truth, will disappear. Conversely, from the point of view of science this time, 'true' religion or 'properly understood' religion cannot conflict with 'real' science. In this case, it suffices to propose a 'good' and 'proper' interpretation of the Scriptures to resolve a conflict that was only apparent. This was the type of argument developed by Dawson in his reply to Draper cited above and by Galileo in his letter to the Duchess of Lorraine. Even a severe critic like Andrew Dickson White, a profound Christian, was convinced that the truths of God 'written upon the human heart' could not contradict those truths 'written upon a fossil whose poor life ebbed forth millions of years ago'.[43] He also believed that there was no real conflict between science and religion and that the battlefield that he described in his book was the effect of the ignorance of theologians who sought to limit the freedom of science. Already in the middle of the nineteenth century, the rationalist philosopher Ernest

Renan had perfectly understood the sophistic logic behind that kind of argument: 'No doubt, truth not admitting of self-contradiction, one would be bound to conclude that *sound science* could not very well contradict revelation. But seeing that the latter is infallible and more clear, if science appears to contradict it, one will conclude that it is not *sound science* and impose silence on its objections'.[44]

The second type of argument is represented here by Cardinal Newman, according to whom the conflict appeared to be the result of a science that had exceeded the limits of its proper domain and in so doing encroached on theology and religion. Conversely, he also admitted that this could also go the other way with theology trespassing on scientific grounds and thus overriding its own limits. According to Newman, to avoid false conflict, each science should therefore stick to its method and its object and remain on its own ground. It is worth noting in passing that the much discussed 'NOMA' (Nonoverlapping Magisteria) principle proposed by the naturalist Stephen Jay Gould as a solution to the conflicts between science and religion, falls under the same general rhetorical form and thus suffers the same limitations.[45] This idea of the 'limits' of science has not convinced the many scholars who consider that science has no absolute limits in its search for a natural explanation of phenomena and that the only way to avoid an open conflict with organized religions is for theologies to adapt their discourse to the current state of science. In this sense, the 'limits' of science are like the limit of the horizon: the more one moves towards it, the more it recedes and one can never reach, and even less 'go beyond', the limit of the horizon.

In general, most people would tend to agree with Cardinal Newman in saying that there *should* be no conflict because the two areas do not have the same object. As the physicist and Catholic philosopher of science Pierre Duhem pointed out at the beginning of the twentieth century, 'In fact, between two judgments not having the same terms nor the same objects, there can be neither agreement nor disagreement'.[46] At the other end of the spectrum of religious sensitivities, consider the philosopher Friedrich Nietzsche, who observed in *Human, All Too Human* that, 'in reality, between religions and real science there exists neither relationship nor friendship, nor even enmity; they live on different planets'.[47] For Renan, it was obvious that 'the so-called problem of the agreement between faith and reason, supposing as it does two equal powers which it is necessary to reconcile, is utterly devoid of sense; for in the first case, reason vanishes in the presence of faith, like the finite before the infinite and the strictest orthodoxy has right on its side; in the second case there remains nothing but reason, manifesting itself in various ways, but nevertheless always remaining identical with itself'.[48]

We must, however, distinguish this type of assertion about what *should* be the case, made from a normative point of view – that amounts by this very fact to a pious wish – and the historical reality of the actual relations between science and religion. This distinction between 'what is' and 'what should be', or between *fact* and *norm*, is essential, yet it is still too often forgotten in the many publications devoted to the relationship between science and religion. From the point of view of historical analysis, our task must be to ascertain the existence of conflict and to recognize, when it is the case, its invariant argumentative structure. These conflicts have a fundamentally social character because it is always a question of struggles between social groups defending their symbolic and social interests and seeking to impose their discourse. The stakes are therefore not only epistemological but also socio-political.

The rise of a science–religion 'dialogue'

As our analysis has shown, the theme of 'conflict' was clearly present in the publications of the various authors who addressed the question of the relationship between science and religion since the 1820s. There are, of course, temporal variations related to special circumstances, including the most recent that saw the emergence, in the 1980s, of the theme of a 'dialogue' between the two domains, an expression completely absent, until recently, from public discourse (figure 5.7). This rise of the idea of a dialogue, which appears to have reached its acme in the second half of the 2000s, can be understood as the convergence of several ideological currents.

First of all, as we saw at the end of chapter 3, in 1979, Pope John Paul II officially called upon the Church to forget the long period of conflict in order to establish a new dialogue with science. More generally, Christian religious groups launched publications focusing specifically on these issues, invariably promoting the idea of a 'dialogue'. And even if the promotion of the idea of a profound agreement between science and Christianity obviously existed prior to the Pope's public intervention, the proliferation of these publications began only in the 1980s.

What is probably the first journal devoted to the relationship between the Bible and modern science was created in 1949 under the title of the *Journal of the American Scientific Affiliation*, a publication

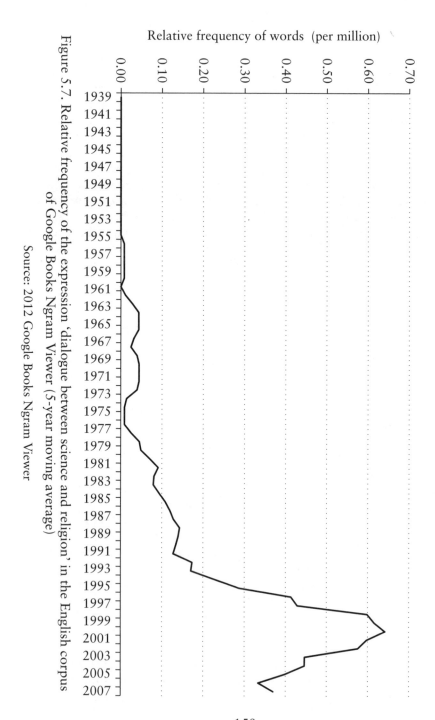

Figure 5.7. Relative frequency of the expression 'dialogue between science and religion' in the English corpus of Google Books Ngram Viewer (5-year moving average)

Source: 2012 Google Books Ngram Viewer

of *The American Scientific Affiliation* founded in 1941, which also published in 1948 a collection of articles on *Modern Science and Christian Faith*.[49] In 1987, the title of the journal was amended to make its objective more explicit: *Perspectives on Science and Christian Faith*.[50] In 1989, another group, Christians in Science, a member of the Evangelical Alliance that benefits from the assistance of the Templeton Foundation (to which we will return) began to publish its own magazine, *Science and Christian Belief*.[51] More recently, *Theology and Science* has been published since 2003, under the guidance of the Center for Theology and the Natural Sciences (CTNS), founded at the beginning of the 1980s by Robert J. Russell, Pastor of the United Church and a man with close links to the Templeton Foundation.[52] The Center organizes conferences on the links between science and theology, again, largely subsidized by the Templeton Foundation. All of the 'big names' associated with the renewal of the 'dialogue' between science and religion have passed through the CTNS: Ian Barbour, Paul Davies, George Ellis, John Polkinghorne and Francisco Ayala.[53] Publishers, sensing a trend and seeking to capture a share of the profits thus generated, have since flooded the market. In 2014, the journal *Science, Culture and Religion*, published by Smith & Franklin Academic Publishing, entered the fray. The journal covers a wider front than theology but still plays on the theme 'science and religion'. Finally, there is the *European Journal of Science and Theology*, created in 2005 and devoted mainly to the Christian points of view on the relations between science and religion. In the Muslim world, the issue of relations between science and religion has also increased its visibility since the 1980s, and there has been an English-language journal dedicated to the topic since 2003. Published under the auspices of the Center for Islamic Studies, founded in Canada in 2000 by the Pakistani chemist Muzaffar Iqbal, it bore the title *Islam & Science* during the first ten years of its existence before becoming *Islamic Sciences* in 2013.[54]

Another important magazine, *Zygon* – a Greek word meaning 'Union' – founded in 1966 by the American Unitarian theologian, Ralph Wendell Burhoe, arose in the context of the counter-culture and the rise of 'New Age' philosophies. It was meant to be (and still tries to be) a forum 'for exploring ways to unite what in modern times has been disconnected – values from knowledge, goodness from truth, religion from science'. The 'New Age' discourse, which seeks to reconcile the most advanced science and the oldest religions has been summarized by the journal's editors who hypothesize that, 'when long-evolved religious wisdom is yoked with significant, recent scientific discoveries about the world and human nature, there results credible expression of basic meaning, values, and moral convictions that provides valid and effective guidance for enhancing human life'.[55] As we will see in the next chapter, this kind of spiritualism mobilizes modern science to disseminate the idea of a 'new alliance' between science and spirituality, while denouncing the former 'paradigm' of a cold and disembodied scientific objectivity.[56] In 1980, the founder of *Zygon* received the Templeton Prize, endowed with a cheque for more than a million dollars, for his 'passionate investigation into the differences and similarities between theology and science'.[57] As we will see later, he was followed in this attribution, starting in the mid-1990s, by many physicists who also suggested links between science, religions and spiritualities.

'Postmodernism' stands as yet another, perhaps more surprising, source of convergence between science and religion. It emerged in the 1980s and one of its central characteristics is the brisk rejection of clear-cut distinctions between categories, the insistence on hybridity, mixture, historicity and the criticism of its opposite: 'essentialism', guilty of negating the (essential?) fluidity of the world. In the postmodern language, 'conflict' disappears, replaced by 'conversation', 'dialogue', 'meeting' and 'exchange' more in tune with the fluidity of things and their shifting 'boundaries'. Postmodernism was also accompanied by a form of conceptual and cultural relativism that seemed 'progressive', but which, in fact, fit easily with the rise of Christian evangelical and Islamic religious fundamentalists that, in turn, resonated with an individualistic and traditionalist view of society – embodied by leaders such as Margaret Thatcher in Britain and Ronald Reagan in the United States.[58] In the United States, for example, this convergence was translated into a revival of the debate on the teaching of the theory of evolution in schools and by a tactical alliance on this issue between Islamic opponents to the theory of evolution and Christian evangelists.[59] The rise of the 'Religious Right' in the United States also corresponds to the election of President Jimmy Carter, avowed Born-Again Christian.[60] *Time* magazine even declared the year 1976 as 'The Year of the Born-Again Christian'.[61]

As figure 5.7 shows, the expression 'dialogue between science and religion' in English-language books increased rapidly in frequency at the beginning of the 1980s, coinciding with the discourse of John Paul II, but did not really take off until the middle of the 1990s, a fact that, in addition to the contributing forces just mentioned, can be explained by the intense activity of the Templeton Foundation.

The Templeton effect on the new 'industry' of the history of the relations between science and religion

The cultural, social and intellectual contexts of the 1980s and 1990s affected historians of science, many of whom adopted (not always consciously) the approach and language of postmodernism. Surveying the recent historiography on the question of the relationship between science and religion, Nicolas Rupke observed that historians now speak 'more of encounters and engagement between the two, less of conflict'.[62]

It is undeniable that the postmodern current, with its criticism of 'essentialism' – more often attributed than demonstrated – and its emphasis on historicism and cultural and cognitive relativism, has marked the lexicon and the arguments of many historians addressing the question of the relationship between science and religion since the beginning of the 1980s. But another factor, rarely mentioned, was the arrival of an important source of funding for those who wanted to show that, after all, there really is (and was) no conflict between science and religion and that, on the contrary, these two areas of knowledge can and must 'dialogue' and 'converse'. This new feature on the landscape, more material than spiritual, was the role played by the Templeton Foundation. A native of Tennessee and devout Presbyterian, John Templeton made a fortune in finance and lived in the tax haven of the Bahamas. Believing that science and religion should come closer to each other for the betterment of society, he created the Templeton Prize in 1972 to reward anyone who contributed to the 'progress' of religion. Having become a British citizen and been promoted to the rank of baron by Margaret Thatcher, he became Sir John Templeton in 1987. The same year, buttressed by that symbolic recognition, he set up the John Templeton Foundation and, ten years later, he founded the Templeton Press. The role of the Foundation became more visible in the mid-1990s when it began to flood the academic world with tens of millions of dollars yearly, dedicated to promoting a new 'dialogue' between science and religion.

The preferred strategy of those promoting public discourse on the relationship between science and religion has always been to foster associations with already-recognized institutions, thus allowing a transfer of credibility. The technique is simple: if great scholars were associated with the Foundation (by receiving a prize, for example), then it certainly must be serious. Thus, in 1996, the Foundation accomplished a masterstroke by convincing the powerful American

153

Association for the Advancement of Science (AAAS) to agree to allow Templeton to sponsor a project entitled 'Dialogue on Science, Ethics and Religion' to the tune of $5 million, a sum that has enabled the programme to be active since 1996. Aware that the rise of fundamentalist Christians in the United States represented a clear and certain danger for the promotion of science and its applications, the leaders of the AAAS cleverly adapted to this trend, thinking they might limit the damage by promoting the idea that science is not opposed to religion. They presented the project as a way to 'facilitate communication between scientific and religious communities'.[63] The Foundation also managed to associate itself with the Royal Society of London in 2004, by donating $280,000 to the Society to organize conferences on the nature of knowledge, a theme just vague enough to allow discussions of possible links with religion and spirituality. As it happened, four of the six speakers were associated with the Foundation.[64] This association generated a debate among British scholars, which significantly dampened the enthusiasm of the Royal Society.[65] In a fitting illustration of the old adage, similar causes produce similar effects, these scholarly societies ultimately acted just as the BAAS had in the 1830s when it too had sought to adapt to the ambient cultural environment, by declaring that science and religion were not in conflict.[66]

In addition to enrolling recognized scientific organizations, another avenue for promoting the ideology of 'dialogue' is to enlist prominent scientific journals. For instance, the journal *Nature*, generally considered 'prestigious' by scientists and which insists that the 'pressure on [their] limited space is severe', nonetheless found ample space (one full page) to promote the idea that 'religion and science can have a true dialogue'. In this article, Kathryn Pritchard, a spokesperson of the Archbishops' Council in the Church of England, publicized their 'Scientists in Congregation' project and explained that it is based on the 'conviction that science and theology can illuminate one another to the benefit of all'.[67] But she never explained precisely how theology could *benefit* science (never defined or distinguished from its social uses) and she reproduced the usual confusions about science as an institution versus the personal religious beliefs of individual scientists. Most importantly, she said nothing about the source of funding for this call for 'true dialogue': namely, two million dollars from the Templeton Foundation.[68] In response to this intrusion of religious discourse in a scientific journal, several distinguished scientists – including the biologist Jerry Coyne and the physicist Alan Sokal – wrote letters criticizing Pritchard's non-arguments, but the

154

editors rejected these letters on the grounds that for lack of space they 'can offer to publish only a very few of the many submissions' they receive.[69] Giving the illusion that scientists' positions on that matter were evenly split, *Nature* published only two letters: one mildly criticizing the idea of dialogue and the other, co-signed by a bishop from the Pontifical Academy of Sciences, that 'wholeheartedly' shared Kathryn Pritchard's view.[70]

In the field of the history of science, we can gain an idea of the influence of the Templeton Foundation by observing that the three most active authors in the criticism of the so-called 'conflict model' of science–religion relations all received the 'The Outstanding Books in Theology and the Natural Sciences Prize', originally funded by Templeton. Awarded, between 1996 and 2000, by the Center for Theology and Natural Sciences, an organization that also benefits from the generosity of the Foundation, this prize is accompanied by a $10,000 cheque.[71] As table 5.1 shows, the list of recipients overlaps with the list of the main players in the new discourse on the relationship between science and religion. John Polkinghorne even appears

---

Table 5.1

Some recipients of the 'Outstanding Books in Theology and the Natural Sciences Prize' financed by the Templeton Foundation (1996–2000)

1996:

Ian Barbour, *Religion in an Age of Science* (1990)

John Hedley Brooke, *Science and Religion: Some Historical Perspectives* (1991)

Paul Davies, *The Mind of God: The Scientific Basis for a Rational World* (1992)

David C. Lindberg, *The Beginnings of Western Science: The European Scientific Tradition in Philosophical, Religious, and Institutional Context, 600 B.C. to A.D. 1450* (1992)

John Polkinghorne, *The Faith of a Physicist: Reflections of a Bottom-Up Thinker* (1994)

1999:

Ronald Numbers, *Darwinism Comes to America* (1998)

John Polkinghorne, *Belief in God in an Age of Science* (1998)

2000:

Max Jammer, *Einstein and Religion* (1999)

Eugene of Aquili and Andrew B. Newberg, *The Mystical Mind: Probing the Biology of Religious Experience* (1999)

---

twice, because he ticks a number of boxes: trained as a physicist – a prestigious discipline – he then turned to theology and became an Anglican priest. He has since become an active promoter of the notion of the harmony between science and religion and also received the much more lucrative Templeton Prize in 2002. A biography, written with the support of the Foundation, was devoted to him in 2011. Entitled *Quantum Leap*, it was subtitled: *How John Polkinghorne Found God in Science and Religion.*[72]

Historians of science could also take advantage of the largesse of the Templeton Foundation in order to offer courses devoted to the relationship between science and religion and theology, to organize conferences and to publish books on this theme. According to the science journalist John Horgan, thanks to grants from the Foundation, almost ninety faculties of medicine in the United States offer courses on the links between health and spirituality.[73] The same is true for university courses on the relations between science and religion such as that offered at the University Alexandre Jean-Cusa of Iasi in Romania, which was able to create a course entitled 'Science, Religion and Philosophy from an Orthodox Christian Perspective', thanks to a $272,000 grant from the Templeton Foundation.[74]

It is quite remarkable to observe (table 5.2) that many of the books that continually repeat the idea that Draper and White were wrong and that there never really was 'constant' or 'inevitable' conflict between science and religion, often owe their existence to the generosity of the Templeton Foundation.

Table 5.2

A few books on the history of science that have received a grant from the Templeton Foundation

William Shea and Mariano Artigas *Galileo in Rome: The Rise and Fall of a Troublesome Genius* (Oxford University Press 2003)

Geoffrey Cantor and Marc Swetlitz (eds) *Jewish Tradition and the Challenge of Darwinism* (University of Chicago Press 2006)

Ronald Numbers (ed.) *Galileo Goes to Jail and Other Myths About Science and Religion* (Harvard University Press 2009)

Thomas Dixon Geoffrey Cantor and Stephen Pumphrey (eds) *Science and Religion: New Historical Perspectives* (Cambridge University Press 2010)

John H. Brooke and Ronald Numbers (eds) *Science and Religion Around the World* (Oxford University Press 2011)

Here again we find the authors who are the linchpin of the new 'industry' of the study of the relationships between science and religion. The thesis denying the existence of conflicts between these two areas became dominant during the 1990s. It is therefore slightly exaggerated to speak of 'New Perspectives' in 2010, as the title of the book edited by Thomas Dixon and his colleagues declares, when the contributors have been singing the same song for at least twenty years. The list in table 5.2 is certainly not exhaustive, and other books may have benefited from Templeton funds in an indirect way. Such is the case with the collection of papers edited (once again) by Lindberg and Numbers, *When Science and Christianity Meet*, which is the fruit of a symposium funded by the AAAS–Templeton programme and held in 1999 at the Center for Theology and Natural Sciences, an organization largely subsidized by Templeton. The University of Chicago Press published the collective work in 2008. To expand beyond the Christian world, the Foundation also helped Geoffrey Cantor, another historian active in the field of science and religion, to organize a symposium on the Jewish tradition and the 'challenge' of Darwinism in 2003, and then to publish the contributions.[75] As we will see in the next chapter, the Foundation did not overlook the Muslim religion and promoted the work of Nidhal Guessoum that aimed to 'reconcile the Koran and modern science'.[76]

The most interesting offering in this fairly homogeneous set of publications was Geoffrey Cantor's contribution to the collective work of 2010 dedicated to John Hedley Brooke, in which he pointed out that the denial of conflicts between science and religion had perhaps gone a little too far and that the time had come to admit that such conflicts exist and have existed! He even went so far as to suggest that such conflicts can exist not only at the social level but also in the minds of individuals.[77] Regardless of how we understand this late awakening to the reality of conflict, it should of course be noted that the Templeton Foundation never imposed its ideas on its beneficiaries who point out, whenever necessary, that they have had complete freedom in their opinions and analyses. But without questioning the good faith of the researchers, we should draw attention to the fact that associations are created by *elective affinities*, which explain that those who accept the money of the Foundation do in fact maintain a discourse that is consistent with the ideology of the funder even though they may have the conviction that they are not directly or indirectly 'influenced'. It seems quite clear that in receiving the Templeton Prize 2007, the philosopher Charles Taylor, known more for his political philosophy than his knowledge of the sciences and their history, did not displease his patron when he declared, without providing a single example, that 'the divorce of natural science and religion has been damaging to both'.[78] One hears here echoes of that Catholic priest, Abbé Maret, who, when promoting a 'Catholic science' in 1834, called for a return to 'the agreement of science and faith, these two forces of the moral world whose fatal division plunged us into a profound intellectual anarchy'.[79] In stark contrast, a Catholic botanist, Brother of the Christian Schools and well versed in the sciences, wrote in 1926, in the wake of the American debate surrounding the teaching of evolution, that at all times and despite the best intentions, concordist temptations 'have harmed religion as well as science itself'.[80]

Finally, it should also be noted that focusing the attention on individual beliefs when important decisions are taken by institutions, is not without calling to mind the kind of agnotology that pushes researchers to busy themselves with all kinds of possible causes of lung cancer except smoking.[81] Focusing the gaze on individual beliefs, incidentally, may render the institutions that censor books and condemn heretics invisible. But one thing is certain: without the enormous financial resources devoted by the Templeton Foundation to the promotion of a 'dialogue' between science and religion, the academic market would have been spared the flood created by this largely redundant literature. It is significant that the historian Ronald Numbers, a central figure in the field of the history of relations between science and religion, who generally objects to making generalizations given the 'complexity' of things, noted in his contribution to the collective book published in his honour, that 'the one generalization [he felt] confident in making [was] that during the past fifteen years or so, the so-called "science and religion dialogue" has spread around the world'.[82] Interestingly, one of the editors of the book, Thomas Dixon, insisted in his introductory chapter that 'we should never stop asking whose interests a particular historical narrative serves and for what purposes it has been constructed. And we should not exempt our own narratives from those searching questions'.[83] Curiously, neither Numbers nor Dixon went so far as to venture an *explanation* for this sudden growth and diffusion of studies devoted to the relations between science and religion. But a modicum of reflexivity suggests that it is not far-fetched to advance that the Templeton Foundation – which by the way 'generously provided a grant towards the costs of the Lancaster conference and of the publication of the book' that contains these contributions[84] – had certainly played an important role in this impressive growth of the academic industry devoted to promoting an ecumenical history of the relations between science and religion.

In sum, the apostles of a 'complex' history of science have in fact avoided studying the existing conflicts. Instead, they have been busying themselves with, for example, the influence of personal religious beliefs on scientific practice, confusing along the way the context of research, always influenced by *personal* choices, and the context of justification, defined by the *collective* norms of the scientific community. By confusing the personal and the collective, they thus give the impression that a 'dialogue' between science and religion is possible. Many analysts also confounded the historians' own positions on the question of possible conflict between science and religion and those taken by the historical actors who, as we have seen, have truly been confronted with conflicting conceptions of the relationship between science and religion in a social context where the issue was the monopoly of the legitimate discourse on nature. Instead of taking seriously what the historical actors have said, they have preferred the promotion of an ecumenism that fits their own social context. These conflicts have, of course, never been 'permanent' or even 'constant', as the straw-man argument repeatedly implies, but certainly *recurrent*. For all conflicts, polemics or controversies eventually run out of steam, and often only resurface some years later, when the situation is appropriate. The case of evolution is exemplary in this regard. As table 5.3 shows, Christian fundamentalist groups, which, in the United States, are opposed to the theory of evolution, have consistently returned to the forefront by using the laws of the

Table 5.3	
Anti-evolution laws adopted in the United States	
Year	State
1923	Oklahoma
1924	Tennessee
1926	Mississippi
1928	Arkansas
1973	Tennessee
1976	Kentucky
1980	Louisiana
1981	Arkansas
1982	Mississippi
1982	Arizona
2008	Louisiana

most conservative states to block the teaching of evolutionary biology or to add the biblical narrative to the school curriculum.

The first phase of legislative success during the 1920s was followed by a resurgence in the 1970s and the beginnings of the 1980s. These laws were invalidated by the Supreme Court of the United States in two important judgments, in 1968 and 1987. The creationist movement has strong organizations that promote their views, like the 'Center for Science and Culture', funded in 1996 'to advance the understanding that human beings and nature are the result of intelligent design rather than a blind and undirected process'. This Center is the emanation of the Discovery Institute, created in 1991 to promote the idea of 'Intelligent Design' and 'creation science'.[85] Its activities led to another important judgment, emanating from a lower court in 2005, declaring that the idea of a 'creation science', was little more than a roundabout way to introduce religion into schools under the guise of science. Be that as it may, in 2006, the Discovery Institute announced that it has 'put over $4 million toward scientific and academic research into evolution and intelligent design'.[86] And as the adoption of a new law in Louisiana in 2008 suggests, creationists and religious fundamentalists will never let go. The Louisiana education law invites teachers, in the name of 'academic freedom' this time – another strategy promoted by the Discovery Institute – to have an open and objective discussion on theories such as evolution, the origins of life and climate change. According to its critics, the Act is in fact aimed at protecting those who teach creationism in science courses.[87]

Since the 1990s, the globalization of communications has allowed American fundamentalist Christian groups to export their combat to Muslim countries. As the historian Ahmad Dallal has noted, the 'Islamic creationist movement seems to aspire to serve as the international arm of the American creationist movement'.[88] In Tunisia, religious fundamentalism even entered physics courses. The physicist Faouzia Farida Charfi tells us that since the beginning of the 1980s she has been 'faced with the resistance of students, expressed with conviction, to the theory of relativity. They claimed that Einstein was wrong, that light propagates with infinite speed'. Without scientific argument, their assertion was based only on 'the belief that light and infinity are the signs of divine power'.[89]

These few examples suffice to show that even the most generous calls for 'dialogue' and 'respect' will hardly prevent the emergence of new conflicts between science and religion.

— 6 —

WHAT IS A 'DIALOGUE' BETWEEN SCIENCE AND RELIGION?

> *The so-called problem of the agreement between faith and reason, supposing as it does two equal powers which it is necessary to reconcile, is utterly devoid of meaning.*
>
> Ernest Renan[1]

The repeated calls for a 'dialogue' between science and religion, whether they came from John Paul II, clergymen, spokespersons for various religions or private organizations such as the Templeton Foundation, have led to a plethora of publications promoting such a 'dialogue'. Rarely, however, do such publications define the exact nature of this 'dialogue'. It is therefore appropriate to ask: What would a real dialogue between science and religion look like?

Dictionary definitions of the word *'dialogue'* tell us that it is derived from the Latin *dialogus*, itself borrowed from the Greek *dialogos*, which refers to philosophical discussions like the Socratic dialogues reported by Plato. This sets us on an interesting track because an essential feature of the platonic dialogues is the presence of argument. Together with an interlocutor, the author of a dialogue exchanges and develops *arguments* that aim to establish theses and thus gain adhesion or consent to those theses. Dictionary definitions also tell us that the word *argument* originally meant 'a reason or set of reasons given in support of an idea, action or theory'.[2] A preliminary characterization of a genuine dialogue would therefore be that it consists of *an exchange of arguments in order to establish a thesis*, a theory or even statement of fact, which seeks to garner consensus among the discussants. As long as there is an exchange of arguments (and therefore of counter-arguments) that advances the state of affairs or the discussion, the dialogue is real. If it stagnates and is limited to unanswered assertions or simple repetition and participants continue to talk past each other, then it is a dialogue of the deaf.[3] For the dialogue to have a chance to engender agreement, the protagonists must have a common ground and speak about the same things. If they are not, then they are unable to understand each other in much the same way as Alexandre Koyré said of Meno and Socrates in his study of Platonic dialogues, when he said that 'their thoughts move on entirely different planes'.[4]

The dialogue as a mode of argument

Armed with this definition, we can ask if a genuine dialogue is possible between science and religion. If religion, or more specifically individual believers of the Christian, Jewish or Muslim religions, or even an adherent of a given spirituality (animistic or otherwise), want to inquire about the status of the material world, they can surely ask a scientist, who will inform them of the state of knowledge on a given topic when the question falls within the purview of the sciences. Thus, to the question 'where do humans come from?', science responds that the most recent knowledge locates their origin in Africa and that our ancestors evolved from animal species that are even older. If by chance the believer responds that the 'methods' of his religion indicate that God created humans directly and that it is impossible that they came from a lower species, is there still a 'dialogue'? It seems doubtful because the scientist will respond that such a *belief* is incompatible with current *knowledge*. This was already Cardinal Newman's position in the middle of the nineteenth century that stated that 'Theology and Science, whether in their respective ideas, or again in their own actual fields, on the whole, are incommunicable, incapable of collision, and needing, at most to be connected, never to be reconciled'.[5]

There cannot, therefore, be a genuine dialogue between science and religion with regard to questions connected to the material world as there is a clear *asymmetry* that makes the exchange unidirectional, going from the scientist to the believer. In general, since the seventeenth century, science has distanced itself from religion. And religions have been led to reinterpret their sacred books in light of the current state of science – more usually, in fact, a state that is *already old*. Here, the flow of knowledge is brief and unidirectional: science explains that some of the interpretations proposed in the sacred books (Bible, Koran, Torah) are not in agreement with the current state of science. In this situation, the choices are limited: (1) adapt the

162

interpretation of religious texts to avoid cognitive conflict, (2) simply ignore that science or, worse, (3) combat it as do creationists and other religious fundamentalists. As will be seen in chapter 7, since the 1980s, the rise of currents of thought that call into question the methods and the results of different sciences in the name of 'local knowledge' and 'spirituality' specific to certain social groups, clearly shows the gap that separates incommensurable conceptions of the world, a gap that makes any true dialogue impossible.

Of course, science still leaves many beliefs intact: praying to God or believing that It created the world *ex nihilo* through love or boredom, does not create problems for science as long as one does not attempt to deduce from these practices and beliefs statements that are incompatible with current scientific knowledge. Thus, to say that a given situation is the effect of a miracle is clearly incompatible with the naturalistic postulates of science. In such a case, science either proposes a natural explanation or simply admits its *present ignorance*, but science will never say that the miracle is *the explanation* of the phenomenon.[6] It is therefore understandable that science does not need a dialogue with religion: as seen in the previous chapters, the sciences have created their own space of dialogue and argument outside of the spaces and territories occupied by religion.

In fact, if we look closely, the calls for a 'dialogue' have always come from religion and their spokespersons, popes, bishops, clergymen and theologians, who seek to use scientific discoveries (in astrophysics, quantum mechanics, neurology, etc.) to lend credibility to their religions. The immense and inescapable credibility of science forces the promoters of religions to seek 'dialogues' with the sciences, as if they needed a 'scientific' foundation to confer credibility to their religious convictions and dogmas.

The apostles of dialogue

We frequently hear that science and religion can be 'complementary'. That seems generous, but what exactly does that mean? First, in order for discourses to be truly 'complementary', they must treat the same object. Thus, it is said that the wave and corpuscular aspects of the electron are 'complementary' because they refer to the behaviour of the same entity, the electron, in different situations (for example: collision or interference). To avoid conflict, to show openness or to simply be accommodating, we might well say that science is 'complementary' to religious beliefs. But this does not establish a dialogue, because the two parties clearly do not address the same objects. Promoters of 'complementarity' have not shown in what sense the two domains *complement* each other, and the least that can be said is that they are *different* domains. So, while the rhetoric of complementarity may be polite, it is also quite empty, probably entertained as a means to avoid overt conflict with religions and their followers.

Another way to justify a dialogue between science and religion is to invoke a supposed 'convergence' of science and religion and invoke the famous 'anthropic principle'. The term may seem learned and impressive to the neophyte, but, in reality, it hides a simple tautology. Although the anthropic principle circulates in popular science works that focus on the relationship between science and religion, it is in no way considered a true 'principle' in physics, like say the 'Pauli principle' or the 'principle of energy conservation'. It was first promoted in the title of a book published by the physicists John D. Barrow and Frank J. Tipler, the latter having subsequently published nothing less than a 'physics of immortality', to which we will return.[7] This so-called 'principle', commented on by many theologians, views the fact that the universe seems to have been arranged so that humanity might emerge to be a profound mystery. This is in fact little more than a reformulation, in apparently more scientific language, of the argument from 'final causes' (or *teleology*) of natural theology. Recall that it was common in the eighteenth century to 'prove' the existence of God by relying on the fact that It had placed the moon at exactly the right place so that the tides were neither too low nor too high.[8] This is of course a tautology, because it is obvious that if the universe (or the Moon) did not have the characteristics required for the existence of life (or of tides, etc.), then there would be no life (or tides). Even the physicist Martin Rees, who nonetheless believes in the weaker 'heuristic' use of the anthropic principle, said that he preferred 'the less pretentious expression "anthropic reasoning"'[9]. This is indeed more precise, because this type of reasoning is intended to simply introduce what serious physicists call 'boundary conditions', that is to say the observed values from which we can then infer the constraints they impose on the possible solutions of the laws of nature. Thus, in order for life (as we know it) to exist on Earth, it is obvious that this planet must not reside too close to the sun. There is nothing mysterious about that and especially nothing upon which to build a serious 'dialogue' over a 'convergence' of science with religion. Questioned on this 'anthropic principle', the Physics Nobel Prize winner Ilya Prigogine, an expert in thermodynamics, responded that it 'means nothing'. Those who draw conclusions on the existence of life from this supposed 'principle' are simply making 'gratuitous assertions' and in so doing relapse into a primitive anthropomorphism that amounts to affirming that 'God created salt water to respond to the needs populations have in salt'.[10] As we will see later in this chapter, publications invoking the 'anthropic principle' or what is called *fine tuning*, that is to say the supposedly exquisite adjustment of the values of the principal constants of nature that make life possible, have made the fortunes of several of the authors near and dear to the Templeton Foundation.

These pseudo-dialogues seem to be a characteristic of the meetings organized by the Templeton Foundation. John Horgan, a journalist who participated in exchanges of this kind at Stanford University in 2003, observed:

> The meeting was supposed to be a dialogue between neuroscientists, such as V.S. Ramachandran, Robert M. Sapolsky, and Antonio R. Damasio, and religious figures, including the theologian Nancy Murphy and the Australian archbishop George Pell. But the dialogue was nominal; *each side listened politely to the other's presentations without really commenting on them.* Several areligious scientists told me privately that they did not want to challenge the beliefs of religious speakers for fear of offending them and the Templeton hosts.[11]

This example seems characteristic of many other so-called dialogues between scientists and religious people.

The empty intersection of two universes of discourse

In sum, the supposed relations between science and religion lack substance and we must conclude instead that the intersection between the two discourses is empty due to the methodological naturalism that excludes any entity transcending nature from the field of science. The National Academy of Sciences, in its efforts to calm down the religious opposition to evolution, expressed this view in its 1998 pamphlet stipulating that 'supernatural constructions are beyond the scope of science' and that 'whether God exists or not is a question about which science is neutral'.[12]

This religious neutrality does not mean that science has no basis in metaphysics. Quite the contrary, this exclusion itself constitutes a founding methodological choice: all sciences must explain the world by natural principles, immanent to nature. There is no absolute proof that this will always work. From this point of view, science is a challenge, a bet on the future: as long as a researcher has not managed to find a natural explanation, he or she must continue to seek it. It is as simple as that. From the point of view of science, 'God', an unspecified 'field of force', a ghost's 'aura' or the 'soul', are only names we give to our ignorance, as Spinoza had already noted in his time. This philosophical principle proposed by a few pre-socratic philosophers or, as they were called at the time *physiologoi*, or natural philosophers, such as Thales, Leucippus and Democritus, is still the basis of current science, even though it is today better equipped, both conceptually and materially through scientific instruments.

Once again, the fact that science, conceived as a collective institution that submits researchers' statements to public criticism, works under naturalistic assumptions, does not mean (or imply) that scientists *as individuals* are immunized against occult beliefs, the basis of which lies precisely in the denial of naturalism. On the contrary, some are strongly tempted to use physical principles to justify the existence of something beyond the material world. At the end of the nineteenth century, for example, the physical concept of ether was invoked to explain the transmission of thought and contact with ghosts – even though these 'facts' had not been clearly established to the satisfaction of the scientific community. The physiologist and Nobel Prize winner Charles Richet, for example, believed in ghosts and tried to communicate with them. These days, quantum physics, deemed mysterious if not incomprehensible, is regularly summoned up to explain paranormal phenomena, even before their existence had been certified.[13]

The uses of a mystico-theological physics

In his classic book *The Formation of the Scientific Mind*, first published in 1938, the French philosopher Gaston Bachelard summarized scientific development in this way: 'scientific progress is at its clearest when it gives up philosophical factors of easy unification such as the creator's unity of action, nature's unity of plan, or logical unity. Indeed, these factors of unity, active though they still were in the pre-scientific thought of the eighteenth century, are never invoked these days'. He added that 'any contemporary scientist wishing to unite cosmology and theology would be regarded as very pretentious'.[14] Half a century later, the ideological context has changed. It now seems that many physicists give the impression of wanting to reunite cosmology and theology or at least arouse religious feelings in readers of popular science works. Such a move may seek to counter the perception of a decline in the tacit support or interest in the sciences and to attempt, moreover, to re-enchant the world that science has otherwise sought to demystify.

The kind of popularization that tends to associate modern science (primarily quantum mechanics, the misunderstanding of which is the well-spring of countless pseudo-scientific ramblings) with various religious and ancient spiritual traditions finds its source in the theosophical circles of the end of the nineteenth and the beginning of the twentieth century.[15] Far from being limited to marginal figures without scientific training, this current of thought, which sought the integration of scientific, philosophical and spiritual knowledge, also attracted recognized physicists. These included Carl Friedrich von Weizsäcker, who published articles in the cultural magazine *Main Currents in Modern Thought*, created in 1940 and edited at the beginning of the 1970s by the physicist Henry Margenau, himself a supporter of the scientific study of extrasensory perception and founder of the reputable journal *Foundations of Physics*.[16] Their ideas did not have much success until 1975 and the publication of the bestseller by the physicist Fritjof Capra, *The Tao of Physics*. In this book that von Weizsäcker had encouraged him to write, Capra attempted to demonstrate that the equations of the quantum theory of fields, an extremely specialized and highly mathematical field of research, resonated with the Taoist, Hindu, Buddhist and Zen texts of ancient wisdom, a syncretic mixture offered up as a form of 'Eastern wisdom' that stood up against the 'western' mechanistic tradition that the 'new physics' somehow rendered obsolete.[17] In a cultural context then hostile to physics, Capra's book aimed explicitly to improve 'the image of science by showing that there is an essential harmony between the spirit of Eastern wisdom and Western science'. He sought to convince his readers that 'modern physics goes far beyond technology' and that 'the way – or *Tao* – of physics can be a path with a heart, a way to spiritual knowledge and self-realization'.[18]

The work of the physicist Ilya Prigogine and the philosopher Isabelle Stengers, *The New Alliance* (1979), also participated in the 'New Age' cultural current that was critical of the mechanistic and disenchanted view of science. Their book was an answer to Jacques Monod's *Chance and Necessity* published in 1970, which advanced a vision of the world in which humans were the product of chance and to a certain extent foreign to nature. To counter this pessimistic message, Prigogine and Stengers declared that, thanks to the thermodynamics of irreversible processes, which study states of matter far from equilibrium and the emergence of order, it was now possible to re-establish contact with a nature that gave rise to both order and the unexpected, thereby re-enchanting the world.[19] The large international symposium 'Science and Conscience' held in Cordoba in 1979 created ample room for this syncretism that sought, according to Henri Atlan, a 'grand scientifico-mystical alliance that was supposed to bring together the rationality of the new physical sciences, quantum mechanics and relativity, a return of the Irrational and the rediscovery of the mystical traditions of the East and the Far East'.[20]

Since the 1990s, however, the tendency to associate God and science has gone well beyond New Age ways of thinking. It is also the result of recognized researchers promoting science by adapting their discourse to a public influenced by a revival of religious militancy. This strategy is not limited to encouraging the public to read popular science books. The Nobel Prize-winning physicist, Leon Lederman, had a political purpose in publishing *The God Particle* (1993). The book appeared in the context of an important debate in the United States surrounding the construction of a particle accelerator, the SSC (Superconducting Super Collider).[21] Physicists were looking for public support to convince elected officials not to cancel the construction of this gigantic apparatus, which would have cost several billion dollars and was already (and always was) over budget. The purpose of the device was the discovery of an elementary particle predicted by theoretical physics, the Higgs boson. The existence of this elementary particle is necessary to explain the origin of the mass of the particles that make up ordinary matter in the universe. Hence the metaphor of 'God particle', in reference to the fact that it created mass and therefore matter. Lederman was clearly playing on strongly held religious sentiments in the United States. Some, more scrupulous, scientists criticized Lederman for pushing the analogy a little too far and thus playing with fire. In reality, he was only following the example of another physicist well known to the general public, Stephen Hawking, though the latter may have in fact been simply joking. In his worldwide bestseller, *A Brief History of Time*, the British physicist claimed that the discovery of the equations for a unified theory of the universe would allow one to know God's thoughts.[22] But even if he were indeed joking, it is significant that he does talk of 'God' in relation to science.

On the Internet site of the online bookstore Amazon, we can view what 'Customers Who Bought This Item Also Bought'. Consider, for example, Frank Tipler's *The Physics of Immortality*, published in 1994, which claims that the immortality of the soul is based on quantum physics. While, unfortunately, we do not know the frequency of the co-purchases associated with books like this, it is nonetheless useful for our purpose which is to identify the kind of readings that are of interest to those who buy the books mentioned here. Thus, those who purchased Tipler's book also purchased (before, after, or at the same time), *The Anthropic Cosmological Principle*, that Tipler had written with John Barrow in 1986, a book advocating a teleological and crypto-creationist vision of nature according to which if the universe exists as it is (with its laws, constants, etc.), it is in order that humans might emerge, which thus indirectly suggests the existence of a Creator.

As noted above, this so-called 'anthropic principle' does not fall within current physics and is in fact a mode of teleological reasoning applied to nature. The idea that the laws of nature 'presuppose' human existence can only be welcomed by the proponents of 'intelligent design', a reformulation in pseudo-scientific terms of a religious concept of the world that is the negation of science, which is, by definition, naturalist. The other co-purchases often associated with Tipler's book demonstrate the strength of the religious metaphor: Paul Davies' *The Fifth Miracle* (1998) and two other titles by this very prolific author: *God and the New Physics* (1983) and *The Mind of God* (1992). All these books present the findings of modern science with titles and allusions suggesting a link between advanced science and a solution to the mysteries of life and the origins of the universe. Lederman's book has often been purchased at the same time as Dick Teresi's *Lost Discoveries* (2002), a book on the ancient roots of modern science that plays on the idea of 'lost' knowledge. We also find among the co-purchases Julian Brown's *Minds, Machines, and the Multiverse: The Quest for the Quantum Computer* (2000) and Hans Moravec's *Robot: Mere Machine to Transcendent Mind* (2000).

The choice of these titles is far from innocent and preys on the allure of mystery, deeply felt by a significant proportion of the population. We should also note that in the United States, where most of these books are sold, more than in France or in England, there is a deep and important evangelist religious current that scientists and their organizations (such as the AAAS that agreed to participate in a Templeton Foundation programme) must take into account. The American biologist Francis Collins, known to the public for his role in the human genome sequencing programme, has also contributed to this accommodation of science to religious sensitivities by publishing *The Language of God* (2007).[23] If the Higgs boson is the particle that

God used to create matter, then it is obvious that the genetic code is the 'language' It used to 'encode' living creatures.

It is striking to note that since the beginning of the 1990s, a multitude of books with eye-catching titles have been published by reputable scientists, most often physicists. While the phenomenon began in the English-speaking world, it is also present to a lesser extent in the francophone world. In France, it is represented by the astrophysicist Trinh Xuan Thuan whose *Secret Melody* (1989, English edition in 1995) also claims that the most recent science resonates with an enchanted vision of nature. Along the same lines, in 1989, the Christmas edition of the popular French weekly, *Le Nouvel Observateur*, published a report on 'God and science' in which one of the articles claiming that there was a convergence between science and faith was signed by the Secretary General of the 'Bena Foundation for the dialogue between science and faith'.[24] Reviving the spirit of the Cordoba Colloquium, Trinh Xuan Thuan joined up with Matthieu Ricard, a media-friendly Buddhist monk, in a series of interviews published in 2001 under the title *The quantum and the lotus: a journey to the frontiers where science and Buddhism meet*. The works of the brothers Igor and Grichka Bogdanov, presented by journalists as falling within the scope of the popularization of science, participates in the same trend. At the beginning of the wave, they had published *God and Science: Towards Metarealism* (1991). They then went on to publish *Before the Big Bang* (2004), followed by nothing less than *The Face of God* (2010) and *God's Thought* (2012), thus following in the footsteps of Paul Davies and his *God and the New Physics* (1983).

In sum, these authors play on ambiguity and mystery, as in the famous American series *The X-Files*, which mixes a little bit of science and a lot of paranormal beliefs. Overall, these books are vaguely reminiscent of *The Morning of the Magicians* (1960), by Louis Pauwels and Pierre Bergier, which presented a series of unusual and allegedly mysterious 'facts', and the works of Robert Charroux whose *Mysteries of the Gods* (1971), *Forgotten Worlds* (1973) and *Masters of the World* (1974) suggested the intervention of extra-terrestrial civilizations to explain 'mysterious' archaeological facts.[25] The major difference here is that today the authors often are respectable scientists and not just simply journalists or science popularizers. But this does not prevent the proliferation of pseudo-science, even when it naïvely aims at making people love science.

The return of natural theology

The wave of popular science conferences and publications with religious overtones has of course been greatly stimulated by the activities of the Templeton Foundation. As its mission is to promote links between science, theology, spirituality and religion, works whose titles combine God and science inevitably attract its attention. In addition to the millions of dollars annually allocated to various projects, it also awards the yearly Templeton Prize, endowed with a grant of a value that is constantly adjusted so that it exceeds that of the Nobel Prize (usually more than a million dollars) as a way to ensure wide media visibility. For in a world where money seems to have become the only measure of 'prestige', relative cash value is important.

Since the mid 1990s, the Foundation's strategy has changed and the winners are not so much individuals associated with conservative and spiritualist ideologies as physicists (often astrophysicists) who have promoted (implicitly or explicitly) links between science and spirituality. Thus, in its early years recipients included figures such as Mother Teresa, the first to be awarded the prize in 1973, television evangelist Billy Graham (1982) or, ten years later, the Russian writer Alexander Solzhenitsyn. Too explicitly associated with the curious idea of the 'progress of religion' listed in the original title of the prize, its designation was amended in 2004 and it now aims to reward those who have made an 'an exceptional contribution to affirming life's spiritual dimension'.[26]

It is therefore hardly surprising to learn that one of the first scientists to receive the Templeton Prize was Paul Davies, in 1995. Very active within the Foundation, Davies is undoubtedly the author of the largest number of books with the words God, science, spirit, miracle, etc., in their titles. Davies is a recognized physicist. He publishes in scholarly journals but has also written many books on popular science that connect science and spirituality in order to attract readers who seek to combine science and transcendence. The fact that he is a university physicist lends credibility to positions that would otherwise be denounced as extravagant.

John Barrow, a member of the United Reformed Church who has done much to propagate the 'anthropic principle', was also rewarded by the Templeton Prize, in 2006. The year before, the physicist Charles Townes, member of the United Church of Christ and a Physics Nobel Prize winner (1964) received the Templeton Prize as a reward for his reflections on 'The Convergence of Science and Religion'.[27] Eschewing no metaphor that might suggest a link between the two, the text announcing Townes' prize even pointed out that Townes had described his discovery of the principle of the maser, which earned him the Nobel Prize, as a 'revelation' and an example of the interaction between 'how' and 'why', between science and religion.[28] Evidence, if such was necessary, that even Nobel Prize winners are not beyond intellectual confusion.

It is interesting to note that Frank Tipler, Barrow's co-author of the reference work on the 'anthropic principle' published in 1986, has never received the Templeton Prize, despite his many books that claim to demonstrate that physics leads to God. It is possible that he went a little too far, thus losing credibility, in publishing *The Physics of Immortality: Modern Cosmology, God and the Resurrection of the Dead* (1994) and *The Physics of Christianity* (2007). The first book appeared at a time when, according to another important actor in these debates, the Quaker astrophysicist George Ellis, the study of the links between science and religion began to gain in importance and social visibility. Ellis had just published an article on the famous 'anthropic principle' and was in the process of promoting the idea of a 'fine tuning' of the main constants of nature, an argument used to suggest the act of a creator. He even claimed that it was difficult, when confronted with this supposed (but in fact nonexistent[29]) adjustment not to use the word *miracle*.[30] Probably fearing that Tipler's book might tarnish the credibility of attempts to bring science and theology together, Ellis went on the attack. He began his review of the book in the British journal *Nature* saying: 'This must be one of the most misleading books ever produced'. Despite the appearances of scholarship and scientific authority, the book was nothing less than a 'masterpiece of pseudoscience'. Ellis concluded his demolition by warning that the reader 'should not make the mistake of believing that this book makes a worthwhile contribution' to the discussion of the relationship between science and theology that was then 'gaining momentum'.[31]

This review was probably the most virulent ever published in the very serious journal *Nature*, which could have simply ignored the book, as it usually does when the contribution is of such poor quality. Its publication thus strongly suggests a settling of accounts in the small community of scientific 'experts' on the relations between science and religion intended to dismiss the slightly more delirious ramblings of some scientists on the relationship between God and contemporary physics. Ten years later, in 2004, Ellis also received the Templeton Prize for his research on parallel universes, the functioning of the human mind, the evolution of complexity, as well as 'the *inter-section* of these issues *with areas beyond the boundaries of science*'.[32] Compared to Tipler, who blithely went beyond the 'boundaries of science', we must believe that Ellis did so in a more subtle way and especially in a way that was more acceptable to the leaders of the Foundation.

In fact, in order to achieve its objective of drawing science, theology and religion closer together, the Templeton Foundation must maintain an aura of credibility and therefore associate itself as closely as possible with prestigious researchers. Thus, it could have been predicted that the British astrophysicist, Martin Rees, president of the Royal Society of London and *Royal Astronomer* would eventually receive the Templeton Prize following his publication of a number of popular science works. The publications began in 1989 with *Cosmic Coincidences: Dark Matter, Mankind, and Anthropic Cosmology.* Then came the subtly religious *Before the Beginning: Our Universe and Others* (1997) and *Just Six Numbers: The Deep Forces that Shape the Universe* (2000) that also suggested the 'fine tuning' of the constants of nature. After speculating on parallel universes, which, by definition, lie well beyond the scope of experiment, Rees plunged, in 2003, into catastrophism with *Our Final Century: Will the Human Race Survive the Twenty-first Century?* In a macabre attempt to render the book even more alarming, the American edition was entitled *Our Final Hour?* The book wondered, among other things, whether our destiny has a cosmic significance.

The author's publishing itinerary seemed to clearly correspond to the requirements of the Templeton Prize. In 2011, the announcement was made: Sir Martin Rees was the happy recipient of the Templeton Prize for his research that 'has contributed to the understanding of the origin and nature of the universe'. Although the formulation avoided the words *religion* or *spirituality*, the announcement was rather poorly received by the many scientists who understood that the purpose of the Templeton Prize was to associate a prestigious name with the activities of a foundation that creates confusion between science and religion. Thus, Jerry Coyne, Professor of Ecology and Evolution at the University of Chicago, published an article denouncing the manoeuvre, reminding readers that 'Templeton plies its enormous wealth with a single aim: to give credibility to religion by blurring its well-demarcated border with science.' He understood that Rees was a 'prize catch' for the Foundation, because 'he's not only a professor, but a Baron, a former president of the Royal Society, master of Trinity College Cambridge and astronomer royal'.[33]

173

Finally, in an effort to expand beyond the anglophone world a little and to increase its visibility and activity in the francophone world, the Foundation has found in the French physicist and philosopher of science, Bernard d'Espagnat, an ideal candidate. His concept of 'veiled reality' subtly suggests that God may well be behind the veil. He received the Templeton Prize in 2009.[34]

In her thorough investigation of the Templeton Foundation, the British journalist Sunny Bains highlighted the fact that most of the recipients had first been members of the Advisory Council of the Foundation in a way reminiscent of David Lodge's 'Small World'.[35] D'Espagnat, for example, was pretty well known around the Foundation before receiving his award. He had contributed to a collection of essays entitled *Science and the Search for Meaning* edited by Jean Staune, who in 1995 founded the Interdisciplinary University of Paris (UIP), an organization devoted to promoting a reconciliation between science and religion.[36] Published in 2006 under the aegis of the Templeton Press, this book had first been published in French, the previous year, as *La Science en quête de sens*.[37] In addition to the contributions of Staune and d'Espagnat, the book contained a chapter on 'Science and Buddhism' by the astrophysicist Trinh Xuan Thuan as well as papers by three other recipients of the Templeton Prize. The book was the result of public and private meetings held in collaboration, once again, with the Center for Theology and Natural Sciences in Berkeley.[38] Staune and his Interdisciplinary University of Paris have also received millions of dollars from the Foundation to amplify public discourse on the relationship between science, religion and spirituality, including a $1.6 million grant for a 2004–7 project entitled 'Science and Religion in Islam'.[39] UIP also received $2 million, between 2004 and 2009, to work with the International Center for Transdiciplinary Research and the Association for a Dialogue Between Science and Religion in Romania on a project to 'establish the field of Science and Theology in the academic and media landscape of Romania and [...] to promote its growth in neighbouring Orthodox countries'.[40] Finally, for the period 2011–14, the UIP was associated with another project costing more than $800,000 and entitled 'Science and Islam, An Educational Approach'.[41] One of the results of these investments was Nidhal Guessoum's *Islam's Quantum Question* that sought to 'reconcile the Koran and modern science', much as 'Catholic science' had tried to do, unsuccessfully, a century earlier.[42] The International Society for Science and Religion, based in Cambridge, England, also took advantage of the Templeton Foundation's financial generosity and in 2007 received nearly $2 million to distribute books on the theme 'Science and Religion' throughout the world and 'particularly in India, China and Eastern Europe'.[43]

To ensure that its message was widely disseminated, the Foundation also created, with the participation of the University of Cambridge, a training programme for journalists writing on the relationship between science and religion. Between 2004 and 2010, $6 million was granted to this programme held in Cambridge, England, and New York and that targeted journalists.[44] It consisted of a series of conferences by researchers and intellectuals chosen by the organizers, which included the usual Templeton programme advocates and prize winners like John Barrow and John Polkinghorne as well as the philosopher Tariq Ramadan, an important media personality on issues affecting Islam – to show that Templeton does not limit itself to the promotion of Christianity.[45] The conferences ended with an oral (but publishable) presentation by seminar participants. Readers can easily gather that it was rather unlikely that these presentations would be critical of the idea of a 'dialogue' between science and religion. Invited journalists received a $15,000 'stipend' for participation and additional funds to reimburse transportation and lodging.[46] In Great Britain and the United States, journalists associated with most of the major newspapers and magazines were invited, a list that included *The Washington Post*, *The Los Angeles Times*, *Discover*, *The New Republic*, *The New Scientist*, *Nature*, *Time*, *The Chronicle of Higher Education*, *Scientific American* and the *BBC World Service*. Francophone newspapers seem to have been spared.

For the programme 'Science in Dialogue' alone, the Foundation invested $56 million in 56 different projects between 1996 and 2013, to which must be added the other millions paid to journalists. With such amounts, which are astronomical when compared to the paltry sums normally allotted to research in the human sciences, we should not be surprised that the theme of a largely defensive 'dialogue' between science and religion has invaded not only public space but also the academic world.

Despite the fact that, in the middle of the nineteenth century, Cardinal Newman had already noted the doubtful utility of natural theology for bolstering the faith of believers, the resurgence of the theme of a 'dialogue' between science and religion has led to the rebirth of the famous *Boyle Lectures*. Essentially pronounced dead at the end of the nineteenth century, they reappeared at the beginning of the 2000s. Given in the same churches, St Paul and St. Mary-le-Bow, supported this time by the Anglican pastor Michael Byrne, they focus on the relationship between science and religion and no longer on the now archaic critique of atheism – although this theme resurfaced in 2014, in response to the rise of the 'new atheism' movement, itself a reaction to the resurgence of religion in the public space since the 1980s. Of the eleven recent *Boyle Lectures*, four have focused on Darwin and evolution and reiterated natural theology theses with titles like 'How Evolution Discovers the Song of Creation'. Unsurprisingly, among the speakers we find the usual promotors of the dialogue between science and religion. Thus, the Reverend John Polkinghorne, physicist and theologian and recipient of the Templeton Prize, as noted above, commented on Keith Ward's 2009 Lecture on 'Misusing Darwin: the Materialist Conspiracy in Evolutionary Biology', and then gave his own lecture in 2013 on the question of the 'dialogue between science and religion'. Two other Templeton prize winners participated in the 2007 *Boyle Lectures* that featured John Barrow on 'cosmology and ultimate ends' and comments by the (then) future Templeton prize winner, astrophysicist Martin Rees. Historians have also been called upon to contribute to the conferences, and we find John Hedley Brooke, who, in 2010, lectured on the legacy of Robert Boyle, followed with a commentary by his colleague Geoffrey Cantor.[47]

The impossible rationalization of faith

At the beginning of the twenty-first century, Pope Benedict XVI remained convinced that Catholicism must not be lived on a voluntarist and subjective basis but rather on a rational (and therefore somehow objective) basis. He explicitly rejected the voluntarist doctrine of the Franciscan Duns Scotus, who opposed to the 'intellectualism of Augustine and Thomas' a God whose decisions 'remain eternally unattainable and hidden'.[48] Against this view, the Pope advocated a return to the rationalist conception of faith embodied in Thomas Aquinas' doctrine, which dovetailed with the new natural theology. While the latter had been able to build a rational system with the backing of the science of his time, based on the philosophy of Aristotle, this ultimately coherent system collapsed in the seventeenth century with the advent of modern science and the dissemination of atomism. From then on, science was no longer in harmony with the dogmas of the Christian Church, which, as we have seen, rather than adapt to the new sciences, preferred to deny their relevance and cling to its dogmas. Even the science of history, which historicized the Bible and made it just another document written by humans, had a lengthy wait before gaining legitimacy within the Roman Church. And that says nothing about the possibility of historicizing the Koran, a path fraught with pitfalls for those who dare to venture down that road.[49]

The systematic application of reason to religious doctrines and their sacred texts could only lead to questions concerning the plausibility and especially the rationality of those beliefs in heaven and beliefs in a personal 'good and merciful' God who dictated his laws to 'prophets'. A crucial step in this interrogation of the rationality of a God of mercy occurred in the middle of the eighteenth century, when European rationalism reached its apex.

The earthquake in Lisbon on 1 November 1755 killed tens of thousands of people, some even as they prayed in church on this day of Ascension. The cataclysm pushed Voltaire to write a long poem 'On the Disaster of Lisbon'. The following excerpt clearly raises the insoluble question of the justification of evil (or rather its impossible justification):

> *But how conceive a God supremely good,*
> *Who heaps his favours on the sons he loves,*
> *Yet scatters evil with as large a hand?*
> *What eye can pierce the depth of his designs?*
> *From that all-perfect Being came not ill:*
> *And came it from no other, for he's lord:*
> *Yet it exists. O stern and numbing truth!*
> *O wondrous mingling of diversities!*[50]

The same cause producing the same effect, the devastation produced by the tsunami in the Indian Ocean on 26 December 2004 (the day after Christmas), which killed more than 200,000 people, has led some believers to ask about the actions (or inactions) of their God. Thus, the Archbishop of Canterbury, the head of the Anglican Church, admitted that such a tragedy 'has made [him] question God's existence'.[51] But as it is better to look on the good side of bad things, some have preferred to see divine action in the international solidarity that followed the disaster, as if a humanist conception of social life presupposed faith in God. Finally, the Muslim God seems hardly more humane towards his supporters, in view of the hundreds of people who perish regularly during the annual pilgrimage to Mecca. The disaster that occurred on 24 September 2015 cost the lives of more than 2,000 faithful. Curiously, these deaths occur most often during the ritual of the stoning of Satan, in the valley of Mina, in the west of Saudi Arabia. As the French daily *Le monde* explained: 'the ritual consists of throwing seven stones on the first day of Eid el-Kébir at a large stone slab representing Satan, and 21 stones the next day or the day after on three other large, medium and small slabs'. The journalist added that 'of the seven major accidents that have plunged the pilgrimage into mourning since 1990, six have occurred during this ritual, the last dating back to January 2006 when 364 pilgrims perished in a scuffle at Mina'.[52] We would need a new Voltaire to express in a poetic manner the absurdity of such events.

Does reason have limits?

Benedict XVI clearly understood that the advent of modern science has had important effects on the relationship between faith and reason. In his famous conference at Regensburg on 12 September 2006, which the media have wrongly reduced to a simple attack on the Muslim faith, while it was in fact essentially devoted to the question of the relationship between faith and reason, the Pope affirmed that he was convinced that by excluding the question of God from the domain of reason, the West had gone astray.[53] In his view, the Kantian critique was the hallmark of a line of thought that was subsequently radicalized by the natural sciences. This movement had led to a 'reduction of the radius of science and reason, one which needs to be questioned'. Because, if science were reduced to its refutable and positive content (the Pope referred here to the philosopher Karl Popper) then,

> The specifically human questions about our origin and destiny, the questions raised by religion and ethics, then have no place within the purview of collective reason as defined by 'science', so understood, and must thus be relegated to the realm of the subjective. The subject then decides, on the basis of his experiences, what he considers tenable in matters of religion, and the subjective 'conscience' becomes the sole arbiter of what is ethical. In this way, though, ethics and religion lose their power to create a community and become a completely personal matter. This is a dangerous state of affairs for humanity, as we see from the disturbing pathologies of religion and reason, which necessarily erupt when reason is so reduced that questions of religion and ethics no longer concern it.[54]

Not seeing any possible basis for ethics and morality other than a reason illuminated by God, the Pope affirmed in a peremptory way that 'attempts to construct an ethic from the rules of evolution or from psychology and sociology, end up being simply inadequate'. Worse still: 'a reason which is deaf to the divine and which relegates religion into the realm of subcultures is incapable of entering into the dialogue of cultures'. Why would a purely secular ethics be insufficient? Why would a dialogue between cultures be impossible for those who have not heard from the Divine? Inquisitive readers received no answers and were forced to settle for these lapidary assertions that resemble injunctions to follow the Pope on the only enlightened path: his, and that of his Church.

Seeking to sound more reassuring, Benedict XVI declared that 'the intention here is not one of retrenchment or negative criticism, but of broadening our concept of reason and its application. While we rejoice in the new possibilities open to humanity, we also see the dangers arising from these possibilities and we must ask ourselves how we can overcome them'. However, the problem is simple: if science has put faith to the side, it is because reason cannot provide a real foundation to faith, as many Catholics at least since Duns Scotus have understood. The trick is therefore to issue a doctrinaire *fiat* that links faith and reason while subtly submitting reason to faith, as if the latter alone could avoid the slippages of a reason left to itself. The only way to control the threats posed to humanity by the new feats of science would be, for the Pope, 'if reason and faith come together in a new way, if we overcome the self-imposed limitation of reason to the empirically falsifiable, and if we once more disclose its vast horizons'. Similarly, in response to the question 'Why this has to be so?' – a question justly evicted from the field of science, which simply accepts 'the rational structure of matter' – the Pope called upon the natural sciences to open themselves up 'to other modes and planes of thought – to philosophy and theology'. In sum, he proposed a return to a pre-Kantian world, linking philosophy and theology without warning, as if it was obvious, two centuries after Kant had defended the autonomy of philosophy vis-à-vis theology in his famous text of 1798 on *The Conflict of the Faculties* (discussed in chapter 1).

This is, in some ways, the new avatar of a concordism that always seeks to unite science and faith, knowing that the former is in many contexts more credible than the latter. But instead of the usual attempts to interpret contemporary science in a way that reinforces Catholic theology, the Pope made a direct request to submit reason to the limits imposed by faith. He thus attempted to build a time machine to travel to a past that is fortunately long gone. And if it is true, as some like to proclaim, that reason has limits and that it 'should now be placed under surveillance', as the Pope said, the

179

problem is that we are never told where exactly this limit is located and especially *who* must patrol the borders.[55] For Benedict XVI, the implied answer is clear: faith should monitor reason, just as the latter should monitor faith.

In fact, this superficial symmetry hides a fundamental asymmetry. While adhesion to a religion is an individual and private act, reason is public and accessible to all normally constituted human beings. In this sense, reason is universal and democratic, while faith is private and autocratic. If rational discussion may eventually lead to a consensus, it is difficult to see how faith can create consensus given that there are many religions in a perpetual state of mutual denunciation. Once again, only woolly thinking that plays on the ambiguity of the words *dialogue* and *opening* could possibly portray as a constructive exchange what is ultimately an attempt to impose an arbitrary religious regime on reason in the name of ill-defined 'limits'. In fact, the Pope's call for a 'dialogue' is little more than a poorly disguised demand for the submission of reason to faith.

Interestingly, the new Pope Francis seems more sensitive to the voluntarist dispositions of Duns Scotus than his predecessor. In his recent foray into climate science in his encyclical *Laudato si'*, Pope Francis repeats the idea that 'science and religion, with their distinctive approaches to understanding reality, can enter into an intense dialogue fruitful for both'. However, despite this ritual affirmation, the whole structure of his text suggest less a dialogue with science than a different kind of answer to the ecological crisis. He says that 'more than in ideas or concepts as such' he is interested in how 'spirituality can motivate us to a more passionate concern for the protection of our world'. For 'the ecological crisis is also a summons to profound interior conversion'.[56] So we do not seem here confronted with 'dialogue' in the sense that we have been critiquing in this chapter and that the more rationalist Benedict XVI was promoting. All the text rather subtly suggests that religion here works alongside, almost as a 'handmaiden' to science (to reverse the famous formula), after the latter has done its work, with the goal of changing collective mentalities (through individual conversions) so as to make life on Earth sustainable. The defence of the autonomy of science does not oppose such division of labour, which leaves to science its total freedom of naturalistic inquiry.

— 7 —

BELIEF VERSUS SCIENCE

Knowledge is of a harder stuff than faith, so that when they collide the latter is shattered.

Schopenhauer[1]

'The Glorious Thirty' (1945–1975) marked the culmination of the belief in progress through scientific development. Since the middle of the 1970s we have witnessed what the political scientist Gilles Kepel has called 'the revenge of God', a global movement accompanied by a questioning of the authority previously granted to science and scientists.[2] One can view this as a form of neo-romanticism, comparable to the movement that followed the Enlightenment at the end of the eighteenth century. It has, for example, become commonplace to oppose the power of 'natural' healing of the human body – assisted by some form of spirituality – to the various 'chemical' products of modern medicine, or even 'spirituality' and 'traditional and ancestral knowledge' to 'western' science, reduced in this way to one viewpoint among others. The rise of fundamentalist religious groups who pressure civil authorities entails a marginalization of science. While science and the rationalism it embodies, aim at – without always attaining – the universal as a regulating principle, groups that refuse science in the name of their beliefs ground themselves in a particularist vision of the world that highlights their differences and their cultural and religious singularities.

The antagonism between particularism and universalism is visible in the conflicts that have opposed, since the 1980s, archaeologists and anthropologists to various religious groups or aboriginal peoples who demand in many countries – including Israel, Australia and the United States – the exclusive control of human bones found on

181

their territories, even when the bones date back thousands of years.[3] In the wake of the rise of aboriginal rights claims during the 1960s and 1970s, indigenous peoples have claimed ancestral rights over their lands and have defined themselves in terms of their singular cultures and spiritualities. This has generally resulted in the adoption of laws governing archaeological research and the management of the objects discovered on the territories occupied or claimed by such groups.

In Australia, for example, in the middle of the 1980s, representatives of Aboriginal Australians obtained restitution of hundreds of human bones aging from 9,000 to 13,000 years old from the University of Melbourne.[4] Ten years later, Aboriginal Tasmanians convinced the government to return to them human bones dating from 12,000 to 17,000 years ago, so as to bury them in their ancestral lands as a 'cure' for the archaeological site that had been desecrated by researchers. For the Aboriginal spokesperson, this choice had nothing to do with science and was consistent with their traditions and spiritual values.[5] At the same time, in response to pressures from ultra-orthodox Jews in Israel, all human bones that had been the responsibility of the Antiquities Authority were turned over to the Ministry of Religious Affairs. For the Director of Antiquities, Amir Drori, this transfer signified 'the end of anthropology in Israel'. But, according to the Director of the Society for the Protection of the Sanctity of the Dead, Rabbi Meir Rogosnitzky, in the Jewish religion, the body and the soul remain connected after death and so scientific research did not justify the 'traumatic violation' produced by the investigations carried out on the bones. Fortunately for the scientists who focus on Neanderthals and the first *Homo sapiens*, orthodox Jews were not interested in the remains older than 5,000 years as they believed humanity began only after this date.[6]

As the journalist Virginia Morell noted, since the 1980s, 'across the globe, archaeologists and anthropologists are making the unhappy discovery that governments are giving cultural traditions and religious beliefs higher priority than scientific inquiry'.[7] Although local electoral and political considerations have influenced these decisions, they have provided the legal bases for groups who value their religious and ideological beliefs above scientific research. These particularisms are a direct challenge to anthropologists who consider that the bones belong to the universal history of mankind and not to one group in particular.[8] A 1979 report prepared by the Association for Physical Anthropology in Canada clearly expresses this universalist conception of science: 'to impede or to curtail archaeological, medical and forensic research on human skeletal individuals because of the religious views of some individuals requires that the vast majority of humanity, including Canada's native peoples, be deprived of the benefits that scholarly research on the dead can offer the living'.[9]

On the one side, archaeologists, prehistorians and anthropologists seek to *objectify* the bones in order to learn more about the human migrations that have marked the Earth's populations. To do so, they must treat the bones as artefacts that they can manipulate and process to extract information to answer questions such as: What was the composition of their DNA?, What did they eat?, What diseases did they have?, etc. All of this is based on the idea that humans have been transformed over time and through their geographical movement, from the first *Homo sapiens* up to the present.

On the other side, some religious groups and indigenous peoples reject the idea of *objectifying* the bones discovered, and consider them 'sacred'. They thus want to prohibit their handling and modification in order to gain knowledge. The question of their actual origins as confirmed by empirical data submitted to scientific criticism is without any real significance for them. For creationists, for example, human beings descended directly from Adam and Eve approximately 6,000 years ago. For several Native American tribes, their ancestors have always lived on their current territory. They thus reject the idea of a migration from Eurasia (and, at the outset, Africa) via the Bering Strait or the Pacific Ocean that occurred more than 12,000 years ago, a hypothesis confirmed by many decades of archaeological excavations and discoveries of human bones.

The 1996 discovery of 'Kennewick Man' has since strengthened the theory of a migration route via the Pacific Ocean. The controversy surrounding this complete skeleton dating from approximately 9,000 years ago, which has pitted American anthropologists against a group of Native Americans living in the region (the southern part of the state of Washington), perfectly illustrates a fundamental truth enunciated by the sociologist Max Weber at the beginning of the twentieth century: 'it should be remembered that the belief in the value of scientific truth is the product of certain cultures and is not a product of man's original nature'.[10] As we will see, the legal debate over the rights to the remains of Kennewick Man truly expressed a conflict of paradigms that define two different global conceptions of the world, to use the philosopher Thomas Kuhn's terms. And it was judges who, to settle the dispute, were led to practise epistemology.

183

A confrontation of two cultures

In the United States, Native American claims from the 1970s concerning the desecration of their cemeteries led the Federal Congress
to pass the Native American Grave Protection and Repatriation Act,
better known by its acronym NAGPRA, in 1990.[11] The law stipulated
that if human bones were found on indigenous territories, they must,
if claimed by a group that has a cultural affiliation with the bones,
be returned to that group. This Act aimed first to ensure that when
human remains from ancient cemeteries were discovered during
excavations, they were to be treated with respect and handed over
to representatives of the Native Americans still living on the territory
and with which the remains had the clearest 'cultural affiliation'.[12] At
the time, several states adopted similar laws.

The consequences of these laws for the world of research were
first felt in the mid-1990s, when the federal government ordered the
5,000 institutions that it subsidized to develop a list of various funeral
artefacts and human remains in their possession and make that list
known to the peoples concerned. The various Native American
groups could thus claim the return of objects that they considered
sacred or culturally important.[13] Unlike the orthodox Jews of Israel
who did not claim bones older than 5,000 years, considering them
beyond their civilization, Native Americans in the United States, as
one of their representatives noted, 'don't accept any artificial cut-off
date set by scientists to separate us from our ancestors'.[14] In Idaho,
on the basis of a law similar to NAGPRA adopted by that State in
1992, a female skeleton more than 10,000 years old was reburied by
the Bannocks, three years after its discovery by archaeologists. This
skeleton was one of the best preserved of some twenty-five skeletons
older than 8,500 years found in the United States.[15] Many other
similar bones have since been reburied in several US states and other
cases remain in dispute.[16]

Of course there are cases of collaboration between the two groups
that allow researchers to conduct scientific studies before the bones
are reburied, but these compromises are also challenged by those
who, invoking technical progress, are opposed to the pure and simple
disappearance of scientific evidence.[17] These cases of compromise
also seem to operate only insofar as the anthropologists' conclusions
confirm the points of view already embedded in the culture of the
groups involved. Thus, a representative of the Umatilla, who claimed
Kennewick Man, responded to those who consider that burying the

skeleton would destroy evidence of their own history by saying that they already knew their history, which had been 'transmitted by the elders and religious practices'. In the Umatilla representative's mind, if the skeleton was more than 9,000 years old, it simply confirmed existing oral history according to which the Umatilla had occupied those lands since the beginning of time.[18] Finally, when it exists, collaboration is carried out with groups that do not call into question scientific techniques and therefore accept the scientific paradigm.

The real problem therefore lies in the cases of radical opposition between incommensurable points of view, which underscore the impossibility of a dialogue between a scientific and empirical approach to the past and a mythic conception of the world based only on ancestral culture without regard to the empirical findings made possible by archaeological research. The Umatillas' lawyer summarized the situation quite clearly: 'Tribal members believe that they have simply been here. Their creation stories have them created from the coyote or from other animals that are indigenous to these places and have always been here. And, again, those stories have to be respected'.[19] Of course, 'respect' in this instance means 'accept'. But, as Ernest Renan noted in the middle of the nineteenth century, science is synonymous with rational critique and 'criticism is no respecter of things; it neither stops at mystery or prestige, it breaks every charm, it pulls aside every veil. This power, utterly lacking in reverence casting an unflinching and scrutinizing glance on everything alike, is from its very essence guilty of high treason against the divine and the human'.[20] In other words, any scientific discovery which calls into question strongly held cultural beliefs can only generate conflict and a dialogue of the deaf.[21]

It was in this context of Native American cultural property claims that Kennewick Man was discovered by chance on the shores of the Columbia River near Kennewick, in the State of Washington, at the end of the month of July in 1996. The skeleton immediately became the subject of a coroner's inquest. Applying the usual techniques of dating, the local archaeologist, James Chatters, called in to help identify the skeleton that he believed was only a few hundred years old, submitted a sample of the fifth metacarpal of the left hand to carbon dating.

According to these measures, Kennewick Man was between 8,230 and 9,200 years old.[22] It was the most complete skeleton, the best preserved and one of the oldest ever discovered in the United States. It therefore had great archaeological and anthropological value because it lent support to the thesis that the settlement of North America was older than generally believed and may have followed a route other than the Bering Strait. Above all, the first observations of the skull suggested that it was probably not a direct ancestor of the Native Americans currently living in the north-west of the United States but that it originated in the islands of the Pacific.[23]

Unluckily for the anthropologists, the skeleton was found on federal territory occupied and claimed by Native Americans and managed by the US Army Corps of Engineers. It was therefore subject to NAGPRA. Upon the announcement of the age of Kennewick Man, his remains were immediately claimed by the representatives of the confederation of Umatilla bands to ensure that the bones were returned to the land according to their ancestral customs. Seeking to avoid problems with the representatives of the bands in the region, which included the states of Washington, Idaho and Oregon, the Army Corps of Engineers applied the Act, supported by the federal government. Thus, a few days before the skeleton was transferred for further analysis under the direction of Douglas Owsley in the Department of Physical Anthropology of the Smithsonian Institute, the Corps of Engineers ordered a stop to any further manipulation of the bones.

Confronted with this refusal to give them permission to study the bones and fearing a permanent loss of access to a source of unique data on the history of the American population, Owsley and a group of anthropologists associated with the Universities of Michigan and Oregon requested, in October 1996, that the court of the district of Oregon issue an injunction preventing the Corps of Engineers from returning the skeleton to the claimants. Among Owsley's group was the anthropologist Rob Bonnichsen of Oregon State University who had already suffered the effects of the new Act. He had discovered fragments of hair on an archaeological site in Montana in 1994, hair that was quickly claimed by a local band in an application of NAGPRA. Although he considered that these remains had not been buried in a cemetery, and were therefore not covered by the Act, he was forced to render a portion of the hair to the FBI, charged with analysing the sample to confirm that they were indeed of human origin.[24]

The scientists' legal argument was based on the fact that the federal Act stipulated that it must first be demonstrated that there was a significant 'cultural affiliation' between the found object and the group making the claim. The government, however, believed that any archaeological vestige dating prior to 1492 was by definition affiliated with the Native Americans still living on the land. The anthropologists argued that there was, in fact, no scientific evidence of significant genetic or cultural ties going back more than 9,000 years that could justify the aboriginal claim. On the contrary, only the detailed study of the remains could establish such an affiliation.

When judges become epistemologists

A long legal saga followed which ended in 2004 with the confirmation by a Court of Appeal of a 1997 lower court judgment that found in favour of the scientists. The Court concluded that there was indeed no substantial evidence of genetic or cultural affiliation between the skeleton and the bands claiming the remains. The three judges of the Federal Court of Appeal thus confirmed the analysis of their lower court colleagues that concluded that a period of 8,000 to 9,000 years 'between the life of the Kennewick Man and the present is too long a time to bridge merely with evidence of oral traditions'.[25] On behalf of his colleagues, Judge Gould of the Court of Appeal added that the only argument in favour of a cultural link between present-day Native Americans and Kennewick Man was based on oral history. The experts had, however, demonstrated 'that oral histories change relatively quickly, that oral histories may be based on later observation of geological features and deduction (rather than on the first teller's witnessing ancient events), and that their oral histories might be from a culture or group other than the one to which Kennewick Man belonged'.[26] For these reasons, the judge concluded that NAGPRA did not apply and that the scientists could return to the study of the cranial morphology, DNA, teeth and diet of Kennewick Man, whose bones resided in the Burke Museum of Natural History at the University of Washington in Seattle.[27]

The fundamental issue in this controversy, which went beyond the legal quibbling over the details of the Act whose application was contested, had not escaped Judge Gould:

> From the perspective of the scientists Plaintiffs, this skeleton is an irreplaceable source of information about early New World populations that warrants careful scientific inquiry to advance knowledge of distant times. Yet, from the perspective of the intervenor-Indian tribes the skeleton is that of an ancestor who, according to the tribes' religious and social traditions, should be buried immediately without further testing.[28]

This last point of view, however, was not without consequences for scientific research. As the lawyers of the Pacific Legal Foundation that supported the scientists' cause noted before the Tribunal, 'there simply must come a time at which a skeleton stops being a relative and becomes an artefact. If there is not such a point, all archaeology must come to an end'.[29]

Incommensurate world views

Whereas the Native American bands rejected 'the notion that science is the answer to everything and [that] therefore it should take precedence over the religious rights and beliefs of American citizens',[30] archaeologists and anthropologists considered that there 'really is a right to study these skeletons and learn something from the person'.[31] This vision of the world promotes the right to know by means of science. In archaeology and anthropology, this means manipulating the bones in order to learn, among other things, more about the origins of the American population. The first conception, by contrast, promotes beliefs based on oral traditions and customary practices. The relation between 'knowledge' and 'belief' was in this case reversed, as shown in this assertion of a Lakota band spokesman at the beginning of the Kennewick Man controversy:

> We never asked science to make a determination as to our origins. We know where we came from. We are the descendants of the Buffalo people. They came from inside the earth after supernatural spirits pre-pared this world for humankind to live here. If non-Indians choose to believe they evolved from an ape, so be it. I have yet to come across five Lakotas who believe in science and evolution.[32]

As the anthropologist David H. Thomas noted in his book on the controversy, the Act itself (NAGPRA) did not say how to 'choose between different visions of the world'. The criterion of the 'preponderance of evidence' does not 'resolve the conflicts between scientific and traditional belief systems whose notion of "evidence" may be entirely incompatible'.[33] We are here confronted with the problem that Max Weber had clearly recognized:

> so long as life remains immanent and is interpreted in its own terms, it knows only of an unceasing struggle of these gods with one another. Or speaking directly, the ultimately possible attitudes toward life are irreconcilable, and hence their struggle can never be brought to a final conclusion. Thus it is necessary to make a decisive choice.[34]

There can, therefore, be no genuine dialogue that would make an agreement possible and only a court can then decide in the light of the existing laws. In the case of Kennewick Man, the judges ruled in favour of scientists. They might have decided in favour of the federal government and the Native Americans, which would have led to a different outcome: the bones would have been returned, there would have been a ceremony, and then the human remains, prized by anthropologists, would have been buried, taking their secrets with them. Since 1990, several human remains dating from 7,000 to 10,000 years ago have been repatriated and buried, concealing forever valuable objects of study for archaeologists and anthropologists.[35]

The passage of time does not seem to have bridged the epistemological gap separating these two modes of thought. In autumn 2012, the anthropologist Doug Owsley, who was one of the original members of the group of researchers who had taken the federal government to court, presented the results of research on Kennewick Man to representatives of the Native American bands who still claimed him as their ancestor. The newspapers reported the event as a 'historic first meeting of two very different worlds'.[36] According to Owsley, Kennewick Man had never really inhabited the region and was most likely a fisherman of Polynesian descent. This conclusion was consistent with the characteristics of the skull and the isotopic analyses of the bones that suggested a diet based on marine mammals. These technical details did not convince the band members present who claimed to be uncomfortable and even traumatized when viewing photos exhibiting the bones from different angles. All have since reaffirmed that Kennewick Man was their ancestor and that he should be buried as soon as possible.

But further research brought a surprise and in 2015, a DNA analysis called into question Owsley's conclusions. The new analysis suggested instead that Kennewick Man was genetically related to the Colville band, in the State of Washington, a group that had agreed to participate in the study by providing samples of their DNA.[37] The debate is ongoing. Because other Native American bands have not participated in the research, the authors of the study admit that they cannot say with any certainty that the Colville are the nearest living descendants of Kennewick man because their comparative DNA database is particularly limited for Native American groups in the United States.[38] Despite this caveat, the US Army Corps of Engineers concluded, in light of the genetic analysis, that the remains are Native American and that interested tribes may now 'submit a claim to acquire the skeleton for burial'.[39] But it can be expected that the controversy will continue, all the more so as Owsley considers that it would be premature, in the light of technological progress, to bury these bones which have yet to supply answers to all of the scientists' questions.

The aboriginal leaders who reclaim Kennewick man also say that they will continue to pressure the US Congress to modify NAGPRA to block any future possibility of an interpretation that favours scientists. The issue here turns on but a few words because it would be sufficient to amend the Act by adding the words 'or has been' after the 'is' contained therein to prevent rulings in scientists' favour. This detail is crucial because the Court of Appeal judges have already established that the Act defines the remains as 'Aboriginal' (*Native American*) only if the 'human remains bear some relationship to a presently existing tribe, people, or culture to be considered Native American'.[40] According to the tribunal, the use of the present tense in the Act therefore requires the demonstration of a link with groups currently living on the territory. To eliminate this constraint, it would be sufficient to amend the Act to cover past links. There has been pressure to do so and many scholarly societies have opposed it. Whatever the final results may be, the claims of indigenous peoples remind us that the conditions of possibility for scientific research are not only epistemological and cultural but political as well.[41]

When prayer replaces medicine

Curious minds eager to know the history of the peopling of the Americas and the cultural practices of their inhabitants can rejoice in the victory of the scientists in the case of Kennewick Man that resulted in the 2014 publication of a collective work of more than 600 pages providing new insights on this issue thanks to analyses of the skeleton.[42] One might also think that the harm that would have been caused had the judges ruled in favour of the aboriginal conception of the world would have been more symbolic than real. After all, knowing where the inhabitants of the Earth really come from does not seem to be critically important for everyday life. Stephen Lekson, an American anthropologist involved in the restitution of hundreds of human bones and cultural objects to the Pueblos, even wondered whether archaeology was 'worth the hurt it caused Native peoples', because many discoveries and interpretations of these remains clearly go against their self-image (wars, cannibalism, etc.).[43] We might say the same of astronomy: after all, what good is it to know (unless this is simply a 'belief') that the universe is approximately 15 billion years old or that the Earth rotates, if this causes pain to Christian or Islamic fundamentalists, or aboriginal peoples?[44]

The question is somewhat more pressing, however, when the refusal of knowledge, scientific methods and techniques lead to the death of a child. We often forget that in the United States, in the majority of states, parents may refuse medical treatment for their children by invoking religious reasons. And yet, up until the middle of the 1970s, many parents were prosecuted and sentenced for criminal negligence, with the courts giving priority to the health and well-being of the child – on the basis of scientific knowledge – rather than to the religious beliefs of the parents. Many Jehovah's Witnesses, for example, who refused blood transfusions, lost custody of their children when their lives were put in danger. There are also many other Christian fundamentalists sects – more than twenty, distributed throughout thirty-four American States – that prefer prayer as a mode of healing (*faith healing*) over any medical intervention.

The political weight of fundamentalist sects has increased sufficiently in the United States since the 1970s that the majority of states have succumbed to pressure to pass laws protecting their followers against any prosecution based on the criminal neglect of their children. This political influence has also contributed to the fact that the United States is one of the few countries in the world that has not ratified the United Nations Convention on the Rights of the Child, which gives clear priority to minors over parents when the health of the child is in danger.[45] An important turning point in favour of the priority of religious rights in the United States occurred with the adoption of a federal regulation in 1974 on the protection of the child that allowed parents to replace recognized medical treatment by simple prayer when their choice was based on a sincere religious conviction. This was so even when it had been demonstrated that the medical treatments were infinitely more efficient than prayer. As if it were not immediately evident, many US studies have taken the trouble to show as much 'scientifically' by calculating the increased rate of mortality within fundamentalist sects as compared to that of the general population. Thirty-eight states and the District of Columbia have religious exemptions in their civil codes on child abuse or neglect; fifteen states have religious defences to felony crimes against children; and twelve states have religious defences to misdemeanours against children.[46]

A study published in 1998 revealed that, between 1975 and 1995, at least 172 children died from a lack of adequate care for diseases that are easy to cure, the parents having been convinced that only prayer could bring their children back to health. Among the more easily treatable cases, we find fifty-two infections, including pneumonia (twenty-two) and meningitis (fifteen), and twelve cases of diabetes. There were also fifteen highly curable cases of cancer.[47] Five sects were responsible for more than eighty percent of the deaths: Christian

Science, Church of the First Born, Endtime Ministries, Faith Assembly and Faith Tabernacle.[48] Taking advantage of a broad interpretation of the First Amendment of the US Constitution on the freedom of religion, a spokesperson for these sects claimed that the government cannot impose restrictions on their beliefs even when these entail the death of children, if their decisions are made on the basis of sincere religious beliefs. By comparison, only two children died for religious reasons in Canada and none in the United Kingdom, two countries that do not protect parents who refuse medical treatment for their children for religious reasons.[49]

In 2009, the Supreme Court of Canada upheld the validity of the Manitoba Act for the protection of children that had forced a teenager to undergo a blood transfusion against the will of her parents. The judges concluded that 'the care and protection of children is a pressing and substantial legislative objective that is of sufficient importance to justify limiting a Charter right'.[50] The judgment confirmed as well a generally accepted principle in advanced societies that the State must ensure the well-being of its citizens and that it can intervene when it finds that parents' decisions run counter to this objective.

The public health problems caused by certain religious beliefs gain still greater importance when the refusal to be vaccinated causes epidemics not only in the immediate environment of the religious group in question but also the death of other children because of the consequences associated with contagious diseases such as measles. In 1991, for example, a city as important and developed as Philadelphia was confronted with a surprising measles epidemic. It was found to be caused by parents who, for religious reasons, had refused to vaccinate their children. Five young people belonging to the Faith Tabernacle sect thus died in ten days, which, according to Dr Paul Offit, Director of the Division of Infectious Diseases of the Philadelphia Hospital, was 'the worst measles epidemic in US history'.[51] To regain control of the situation, the health authorities obtained a decision from the court ordering the vaccination of children of parents belonging to the sect, a number that amounted to more than 400 children. According to the judge, the freedom of religion was not absolute and the health of children and that of the population as a whole had precedence.[52]

At the beginning of the twentieth century, during a smallpox epidemic, the Supreme Court of the United States had also upheld a law imposing a fine on those who refused to be vaccinated, even for religious reasons, stating that the freedom provided by the Constitution was not absolute as that would only lead to anarchy and disorder.[53] Judgments like these did not, however, prevent the

192

House of Representatives from passing a law in March 2014 limiting the obligation of citizens to obtain health insurance as required by President Obama's health insurance programme, for those who had a sincerely held religious belief opposed to any medical treatment.[54] For these sects, which comprise hundreds of thousands of people, Jesus and God are, through the intercession of prayer, the only real doctors. In such a context, it is not hard to understand why the infant mortality rates within the fundamentalist sects in the United States who are opposed to medicine are comparable to those of the least developed countries of the world.[55]

The 'western medical paradigm' versus ancestral medicine

The excesses of a world view that opposes science on the basis of religious, ethnic or cultural particularities were tragically illustrated in a November 2014 decision by a lower court judge in Ontario, Canada. The judge refused to apply a decision of the Supreme Court of Canada in a case in which doctors sought to treat a child suffering from a serious, but curable, cancer against the wishes of the parents who believed that prayers could bring about a cure. In a decision that generated much comment in the Canadian media, the judge, himself belonging to the same First Nation as the parents, concluded that the child did not need the protection of the state.[56] He considered that, as members of a First Nation, the parents had exercised their rights, protected by the Canadian Constitution, to practise their own traditional medicine. In addition to the fact that the 'treatment' offered to the child was not derived from aboriginal traditions but from a Florida quack practising 'holistic' medicine in the process of being sued in that State for the illegal practice of medicine, what is striking here is that at no time did the judge really take into account the actual effectiveness of the different treatments. In effect, while the doctors had shown, during their interrogation, that the rate of healing for the type of cancer affecting the young girl (acute lymphoblastic leukaemia) was approximately ninety per cent and that no case of survival without such treatment was known, the judge concluded nonetheless that 'D.H. [the mother]'s decision to pursue traditional medicine for her daughter J.J. *is her aboriginal right.*' He even added that 'such a right *cannot be qualified as a right only if it is proven to work by employing the western medical paradigm.* To do so would be to leave open the opportunity to perpetually erode aboriginal rights'.[57] According to the judge, the parents were not

193

guilty of wrongdoing because they had simply exercised their constitutional rights. Blinded by an abstract legalism oriented towards the safeguarding of aboriginal rights without regard to their nature, the judge suggested in sum that the children of indigenous peoples should continue to receive 'traditional medicine' so as not to see their rights 'perpetually erode', even though the mortality statistics indicated that these 'spiritual' treatments had no efficacy and led to a certain death.

Non-aboriginal peoples like Jehovah's Witnesses, however, may not avail themselves of this constitutional 'right' and can therefore be taken in hand by public health officials and administered life-saving treatments. What the Ontario judge – in a fit of postmodern language – called the 'Western medical paradigm' is in fact the sum total of knowledge that has withstood the various tests and trials to which all medical treatment are submitted. Bleeding, for example, a panacea until the nineteenth century, has been today widely abandoned just like many other 'western' remedies. By opposing a 'traditional' medicine to 'western' medicine, the judge adhered to an extreme form of relativism that refuses to consider even the simple empirical validation and comparison of the effectiveness of treatments. He thus condemned a young aboriginal girl to a certain death, while other children of the same age and suffering from the same potentially fatal disease, but that had not been born members of a First Nation, would have their lives saved thanks to the latest advances in medical science.

Fortunately, pressures on the Department of Justice of the Province of Ontario to appeal this judgment have surprisingly borne fruit. At the end of April 2015, following behind-the-scenes negotiations between the lawyers of the different parties, the judge 'clarified' his decision – in the diplomatic language of lawyers – so that the right to use treatments falling within the scope of traditional medicine 'must remain consistent with the principle that the best interests of the children remain paramount'.[58] Between the months of November 2014 and March 2015, the parents of the child probably also realized that their daughter was going to die and therefore decided to urgently resume the chemotherapy treatments, in combination with 'traditional medicine', a compromise that saves appearances, as adding prayer to proven remedies can do no harm to anyone, while giving the impression that the two practices are somehow 'complementary'.

It is often said that 'ignorance of the law is no excuse'. In the presence of legal judgments with irrevocable consequences, one may ask whether judges and other jurists may with impunity ignore the 'laws' of science, i.e. the confirmed conclusions of science. Just as prayers have not healed the hundreds of children left without medical care by parents blinded by their religious beliefs, we fear that the most radical proponents of the 'ancestral rights' of aboriginal peoples to practise their 'traditional medicine' without taking account of the evolution of knowledge and technology will continue to lead to the loss of the lives of children sacrificed on the altar of particularistic ideologies.

A choice and its consequences

The examples discussed in this chapter graphically illustrate the opposition between the world of science and that of fundamentalist religious beliefs. There are, of course, between these two extremes, many shades of gray. It is nonetheless the case that in practice judges have often had to decide between two contradictory conceptions of the role of science in society. In the case of Kennewick Man, the choice was between accepting the sacred nature of the bones and returning them to the Native Americans, or, on the contrary, treating them as non-sacred artefacts worthy of study and thus entrusting them to anthropologists and archaeologists. In the first case, it would be signing the death warrant of anthropology as a science. In the second, it would build on the cultural importance of the advancement of knowledge on migration and the origins of humanity. Favouring tradition to the detriment of reason, the Ontario judgment would have caused – if the social pressure had not forced the judge to amend his decision – more than the death of a scientific discipline but that of a human being.

These choices are therefore crucial and before promoting the naïve and very 'postmodern' notion of the multiplicity of 'local truths', we must consider that there are consequences that will have to be accepted. Supporting religious fundamentalisms of any nature, beliefs about the magic of prayers or the powers of 'ancestral' and 'natural' medicine and refusing the methods and results of the most rigorously tested science under the pretext that they are 'western', can have tragic consequences. Similarly, prohibiting research on behalf of the religious convictions of certain minority groups, such as research on stem cells, once again has consequences. It may be that this research will ultimately turn out to be futile or even dangerous. But it may also lead to fascinating discoveries and even, some day, to useful applications. As we know, societies are mortal. As instanced in the return of infectious diseases because of rising objections to immunization, it is likely that the next public health disasters will be the result of a surfeit of romantic beliefs rather than too much science and knowledge.

CONCLUSION:
BETTING ON REASON

Even at the price of being considered narrow-minded, one must stand for Reason.

Georges Canguilhem[1]

Science, in its attempts to explain natural and social phenomena through natural causes and not through appeals to supernatural causes, disenchants the world. As we have seen in this book, this naturalist bias has slowly imposed itself in all areas of knowledge. The assumption of a methodological naturalism that is at the basis of science cannot, of course, itself be demonstrated and constitutes in fact a bet on the future. Its acceptance will ultimately depend on the results it produces. It is in large extent validated by its consequences, which include the increase in intelligibility that its adoption produces. And it is truly the exercise of an increasingly systematic secular reason that has led to an improved knowledge and a greater control of our natural environment and its radical transformation.[2] Thus vaccination, despite the religious objections that it continues to raise, has proven its worth for several centuries, just as have the many medical interventions that have saved the lives of children and newborns, who would otherwise have died.

Advances in molecular biology have also shown the fertility of the naturalist postulate deployed in the sciences. The biologist Craig Venter, for example, viewed by many as supremely arrogant, claims that he will one day be able to create life from the synthesis of DNA. We could throw up our hands in horror and say that he is 'playing God', and many have not hesitated to do so.[3] We may also note, somewhat more serenely, that in doing so he has wagered on the future: we will see whether, on the basis of the laws of biochemistry, he succeeds in creating a completely synthetic virus and then a synthetic cell that will reproduce on its own. What is important here is that Venter's materialistic, reductionist and biochemical assumptions allow him to advance research. We could, on the contrary, speak of an 'élan vital', like the spiritualist philosopher Henri Bergson did at the beginning of the twentieth century, or of 'the breath of life' which can come only from the Creator, as do modern creationists, or even declare that life is a sacred object and cannot therefore be studied without constraints. No need then to do biochemistry, but we will thus discover nothing new and especially not the structure of DNA as Watson and Crick did in 1953, a discovery that launched molecular biology as well as the biotechnology industry twenty years later.

Far from being free from any presuppositions, scientific activity is based on metaphysical, ethical and methodological assumptions that are the product of a long history. This normative structure of science, first formalized by the American sociologist Robert K. Merton, has a functional character and relies fundamentally on the postulate of universalism as the regulating principle of scientific activity: knowledge has a universal scope, independent of individual particularities.[4] In other words, Euclid's geometry (ca. 300 BCE) is universal even though Euclid was Greek and probably believed in the power of the local gods. Similarly, Ptolemy's astronomy (90–168 CE) was subsequently developed by Muslim astronomers such as Ibn al-Haytham (965–1040), a great critic of the Ptolemaic model, who thus contributed to the advancement of universal knowledge about the movement of the stars, an advancement that led to the astronomical model of the Catholic Canon Nicolaus Copernicus (1473–1543). According to this conception, there is no 'Aryan' science, as the Nazi ideologues believed, or a 'proletarian' science as their Stalinist counterparts thought, or a 'Catholic' science as some intellectuals dreamed of at the end of the nineteenth century, or even an 'Islamic science', an idea which has arisen since the end of the 1970s thanks to the support of countries such as Saudi Arabia.[5] Another regulating norm requires the collective verification of statements submitted to critical scrutiny. In a word, scientific *objectivity* consists in the *intersubjectivity* of knowledge. Science is, by definition, collective knowledge and stands against personal and subjective beliefs that remain private and non-verifiable by other persons with the appropriate expertise. Religions, like spirituality, due to their irreducible multiplicity and despite the imperialist impetus of some of them, remain particularisms, whereas science aims to overcome these particularisms through empirical verification and logical coherence. Finally, the scientific approach to nature is based on the rationalist postulate that the world is understandable and can be explained by laws immanent to nature. As Ernest Renan clearly stated in the middle of the nineteenth century, science has 'restored to nature that which [was] formerly looked upon as superior to nature'.[6]

The critics of 'Western reason' – who frequently confuse scientism and rationalism – will counter that the 'technosciences' have also created chemical pollution, atomic bombs, nuclear waste, etc. But we must distinguish science from technology. The first is only a way of *explaining phenomena by natural causes*, whereas the second develops artefacts (technology) for civilian or military purposes. That modern science relies on instruments (telescopes, microscopes, etc.) is obvious, but that does not make it a 'technology' and even less the confused and ill-defined hybrid called 'technoscience', a concept that is more polemical than analytically useful.[7] Obviously, we will not see further by turning off the lights. If science, or more exactly, its use by some social groups, has led to destruction, it is only through more science – and not more fasts or prayers – that we will find solutions, to the extent that these solutions depend on scientific insights. Religions and spiritualities may well have their place in terms of influencing ethical personal and collective choices, but the confusion introduced by the whole paradigm of 'dialogue of science and religion' has obscured the distinctiveness of that role, as it has the equally distinct role the sciences can and must play in solving our collective problems. Scientific and technical progress has been made possible because some individuals, believing that the world is knowable, subject to universal laws and not to the whims of the gods, have taken risks and questioned traditional beliefs.

Science is a form of objectivity based on intersubjectivity, in the sense of an agreement between informed individuals who discuss and debate together. It is the possibility of achieving a consensus that makes science an *institution* and that allows us to speak in its name. Exchanges are then possible because researchers implicitly accept a number of rules, including the suggestion that one should seek natural explanations and test them, as far as possible, empirically, i.e. confront them with reality. Without this common agreement, there would be no real discussion possible and points of view would become incommensurable. When Nobel Prize winners in physics or chemistry express themselves on matters that fall within their field of competence, they do not do so in their own name, as if expressing a simple *opinion* or personal *belief*, but in the name of the *knowledge* that has been collectively sanctioned. On the contrary, when scientists use their public fame to pronounce on their personal religious beliefs, as if such views had more value than those of the common citizen, they are in fact usurping their authority and abusing their symbolic power.

Science is objective not only in the sense that, in certain circumstances, it corresponds to reality (without being its simple reflection), but above all in the sense that it aims towards the universal. Scientific theories are dynamic and therefore change depending on new ideas and discoveries. Even the hard core of its various theories (i.e. the most entrenched assumptions of chemistry or physics, for example) sometimes change, at the discretion of scientific revolutions. In sum, to be mistaken and to correct its mistakes are integral parts of the game of science.[8] By contrast, although some of the many religions claim universality, the fundamentally subjective nature of these beliefs makes the idea of a universal religion, in the sense that one can speak of a universal science, an illusion.

To avoid entertaining an epistemological and semantic debate about the multiple meanings of the words *know* and *believe*, we will simply remind the reader of their most widely accepted sense that refers to *knowledge* as collectively validated by generally recognized procedures and to *beliefs* as personal convictions not yet validated collectively.[9] It is always possible to play with words – like many postmodern cynics – by combining them in multiple ways to say that we 'believe we know' or 'know we believe'. This is, however, of little use if one accepts that, beyond some divine gaze overlooking nature, all know-how and all knowledge is the product of a community that exists at a given time in history and remains in principle fallible and thus open to future revision or rebuttal. It is nonetheless the case that we generally accept that the statement 'I *think* that 2 + 2 are 4' is confused, while 'I *know* that 2 + 2 are 4' is clear. Beliefs can be revised if they are not regarded as dogmas, but they differ from knowledge as long as they remain individual or subjective and not collectively validated according to procedures regarded as legitimate that may also change with time in light of the empirical results that they achieve. The history of science has long shown that scientific knowledge evolves, even if not everything can be called into question at any given time. In addition, empirical facts are more stable than the theories that seek to explain them. Those theories, moreover, may themselves change without amending the empirical basis that underwrites them. That material bodies at the surface of the Earth fall remains a fact even if the explanation for their fall has changed from Aristotle to Newton and once again with Einstein. Despite the many nuances that are possible, we need to be wary of falling into confusion and thus we must continue to distinguish knowledge from belief.

In his criticism of reason in 2004, Cardinal Joseph Ratzinger, one year before becoming Pope Benedict XVI, suggested that one should accept 'a necessary form of correlation between reason and faith, reason and religion, destined to a mutual purification and regeneration'.[10] Against this appeal to a 'dialogue' based on the implicit assumption that religious faith can enlighten reason, we must recall that the limits of reason are historically mobile and that it is therefore still only reason, embodied in public debates (scientific, ethical and political), that is able to control what Benedict XVI called the 'pathologies of reason'. Religion cannot oversee reason and judge it: there is no religious superego supervising reason. In a disenchanted world, forever subjected to the 'war of the Gods', as Max Weber said, only science and reason can correct the errors of science and reason.

NOTES

Epigraph

1 Émile Durkheim, *The Division of Labor in Society*, translated by W.D. Halls, New York: The Free Press, 1984, p. 119.

Introduction

1 Ernest Renan, *Œuvres complètes*, tome 1, Paris, Calmann-Lévy, 1947, p. 961; our translation.

2 Frère Marie-Victorin, *Science, culture et nation*, Yves Gingras, ed., Montreal: Boréal, 1996, p. 85.

3 Yves Gingras and Geneviève Caillé, 'Nouvel Âge et rhétorique de la scientificité', *Interface*, vol. 18, no. 2, March–April, 1997, pp. 6–8.

4 Max Weber, *From Max Weber: Essays in Sociology*, translated by H.H. Gerth and C. Wright Mills, New York: Oxford University Press, 1946, pp. 350–1.

5 The literature on this topic is too extensive to be cited in full. For a useful bibliography, see Randy Moore, Mark Decker and Sehoya Cotner, *Chronology of the Evolution-Creation Controversy*, Santa Barbara, CA: Greenwood Press, 2010.

6 David B. Wilson, 'On the Importance of Eliminating *Science* and *Religion* from the History of Science and Religion: The Cases of Oliver Lodge, J.H. Jeans and A.S. Eddington', in Jitse van der Meer (ed.), *Facets of Faith and Science*, vol. 1, Lanham, MD: University Press of America, 1996, pp. 27–47; David N. Livingstone, 'Which Science? Whose Religion?', in John Hedley Brooke and Ronald L. Numbers (eds), *Science and Religion Around the World*, Oxford: Oxford University Press, 2011, pp. 278–96.

7 Pierre Gisel, *Qu'est-ce qu'une religion?*, Paris: Vrin, 2007.

8 Danièle Hervieu-Léger, *La Religion pour mémoire*, Paris: Éditions du Cerf, 1993, p. 119, cited by Gisel, *Qu'est-ce qu'une religion?*, p. 16.

9 Danièle Hervieu-Léger, 'Faut-il définir la religion? Questions préalables à la construction d'une sociologie de la modernité religieuse', *Archives des sciences sociales des religions*, vol. 63, no. 1, 1987, p. 13.

10 George Saliba, *Islamic Science and the Making of the European Renaissance*, Cambridge, MA: MIT Press, 2007, pp. 233–55; Ahmad Dallal, *Islam, Science and the Challenge of History*, New Haven, CT: Yale University Press, 2010.

11 David C. Lindberg and Ronald L. Numbers (eds), *God and Nature: Historical Essays on the Encounter Between Christianity and Science*, Chicago, IL: University of Chicago Press, 1986; David C. Lindberg and Ronald L. Numbers (eds), *When Science and Christianity Meet*, Chicago, IL: University of Chicago Press, 2003.

12 Donald M. Leavitt, 'Darwinism in the Arab World: The Lewis Affair at the Syrian Protestant College', *The Muslim World*, vol. 71, no. 2, 1981, pp. 85–98; Nadia Farag, 'The Lewis Affair and the Fortunes of al-Muqtafat', *Middle Eastern Studies*, vol. 8, no. 1, 1972, pp. 73–83; Marwa Elshakry, 'The Gospel of Science and American Evangelism in Late Ottoman Beirut', *Past & Present*, no. 196, August 2007, pp. 173–214.

13 Stefano Bigliardi, 'The Contemporary Debate on the Harmony Between Islam and Science: Emergence and Challenges of a New Generation', *Social Epistemology*, vol. 28, no. 2, 2014, pp. 167–86.

14 Seung Chul Kim, 'Sunyata and Kokoro: Science-Religion Dialogue in the Japanese Context', *Zygon*, vol. 50, 2015, pp. 155–71.

15 See Daniel Dubuisson, *L'Occident et la Religion. Mythes, science et idéologie*, Brussels: Éditions Complexe, 1998.

16 Examples of this kind of confusion are too numerous for a complete list. See, for example, Michael J. Crowe, 'Astronomy and Religion (1780–1915): Four Case Studies Involving Ideas of Extraterrestrial Life', *Osiris*, vol. 16, 2001, pp. 209–26.

17 John Hedley Brooke, *Science and Religion: Some Historical Perspectives*, Cambridge: Cambridge University Press, 1991, p. 5.

18 Peter Bowler also notes the danger for the complexity thesis of 'degenerating into a welter of case studies of interest only to specialists'. See Peter J. Bowler, 'Complexity in practice', *British Journal for the History of Science*, vol. 44, no. 2, June 2011, p. 279.

19 Albert Einstein, 'Personal God Concept Causes Science-Religion Conflict', *The Science News-Letter*, vol. 38, no. 12 (21 September 1940), pp. 181–2. For an analysis of the way different religions use Einstein, see Max Jammer, *Einstein and Religion*, Princeton, NJ: Princeton University Press, 1999.

20 David C. Lindberg and Ronald L. Numbers, 'Beyond War and Peace: A Reappraisal of the Encounter Between Christianity and Science', *Church History*, vol. 55, 1986, pp. 338–54.

21 Brooke, *Science and Religion*, p. 108.

22 *Ibid.*, pp. 234–5.

23 Richard Olson, *Science and Religion, 1450–1900: From Copernicus to Darwin*, Baltimore, MD: Johns Hopkins University Press, 2004, p. 183.

24 Peter Harrison, *The Territories of Science and Religion*, Chicago, IL: University of Chicago Press, 2015, pp. 172–3. This book is a product of the Gifford Lectures whose objective since their foundation in 1888 has been the promotion of natural theology and knowledge of God (www.giffordlectures. org/); for more details on the history of these Lectures, see Larry Witham, *The Measure of God. Our Century-Long Struggle to Reconcile Science & Religion. The Story of the Giffords Lectures*, New York: HarperSanFrancisco,

2005. Understandably, given this context, the author (consciously or not) minimizes the tensions between science and religion and, like many others, promotes the thesis of the 'myth' of the conflict between science and religion, to which we will return in chapter 5.

25 Maurice A. Finocchiaro, 'That Galileo was imprisoned *and* tortured for advocating Copernicanism', in Ronald L. Numbers (ed.), *Galileo Goes to Jail and Other Myths About Science and Religion*, Cambridge, MA: Harvard University Press, 2009, p. 74. Lindberg and Numbers also write that Galileo 'was never tortured nor imprisoned – *simply silenced*', in Lindberg and Numbers, 'Beyond War and Peace', p. 347, emphasis added. Here they use the frequent trick of combining two different cases in one sentence through an apparently innocuous 'and' which makes of course 'false' the sentence as soon as at least one of the components is false. Since it is obvious to all since at least the end of the nineteenth century that Galileo was not tortured then the sentence is false even though one can consider it legitimate to say that Galileo was indeed incarcerated and thus, in the general sense of the word, 'imprisoned'. Finocchiaro used the same trick in the title of his paper and repeated it on p. 73.

26 See, for example, Galileo *Opere*, vol. 16, letters 3268, 4 March 1636 (p. 400) and 3272, 15 March 1636 (p. 406), and in volume 17 letter 3684, 20 February 1638 (p. 297).

27 http://www.oxforddictionaries.com/definition/english/prison.

28 See, for example, the collection of interviews with scientists, *Le Savant et la Foi*, edited by Jean Delumeau, Paris, Flammarion, coll. Champs, 1991.

29 See, for example, Crowe, 'Astronomy and Religion (1780–1915)', p. 225, and John Hedley Brooke, 'Religious Belief and the Content of the Sciences', *Osiris*, vol. 16, 2001, pp. 3–28.

30 Don O'Leary, *Roman Catholicism and Modern Science: A History*, New York: Continuum, 2007.

31 Pierre Duhem, *The Aim and Structure of Physical Theory*, Princeton, NJ: Princeton University Press, 1954, pp. 38–9.

32 For a critique of the rhetorical nature of the 'new natural theology', see Barbara Herrnstein Smith, *Natural Reflections: Human Cognition and the Nexus of Science and Religion*, New Haven, CT: Yale University Press, 2009, pp. 69–120.

33 Yves Gingras, 'Duns Scot *vs* Thomas d'Aquin: le moment québécois d'un conflit multi-séculaire', *Revue d'histoire de l'Amérique française*, vol. 62, no. 3–4, 2009, pp. 377–406.

34 For an analysis of the numerous theological interpretations of contemporary science, see François Euvé's 'Bulletin' that reviews books on theology and science in the journal *Recherches de science religieuse*, vol. 96, 2008/3, pp. 459–78; vol. 98, 2010/2, pp. 303–19; vol. 100, 2012/2, pp. 295–312; vol. 102, 2014/4, pp. 609–32; see also Alexandre Ganoczy, 'Quelques contributions récentes au dialogue entre sciences de la nature et théologie', *Revue de science religieuse*, vol. 94, 2006/2, pp. 193–214.

35 Stefaan Blancke, Hans Henrik Hjermitslev and Peter J. Kjaergaard (eds), *Creationism in Europe*, Baltimore, MD: Johns Hopkins University Press, 2014.

1. The Theological Limits of the Autonomy of Science

1 Charles Joliet, *L'Esprit de Diderot. Maximes et pensées*, Paris, Balland, 2013, 78.
2 Sébastien Morel, *Christianisme et Philosophie. Les premières confrontations (Ier–VIe siècles)*, Paris: Livre de poche, 2014; Georges Minois, *L'Église et la Science. Histoire d'un malentendu: de Saint Augustin à Galilée*, Paris: Fayard, 1990.
3 Nathalie Raybaud, 'Le logos en terre d'Islam: Averroès contre al-Ghazali', in Laurence Maurines (ed.), *Sciences & Religions. Quelles vérités? Quel Dialogue?*, Paris: Vuibert, 2010, pp. 41–9.
4 Luca Bianchi, *Censure et liberté intellectuelle à l'université de Paris (XIIIe–XIVe siècles)*, Paris: Les Belles Lettres, 1999; Alain de Libera, *Foi et Raison*, Paris: Seuil, 2000.
5 Immanuel Kant, 'The Conflict of the Faculties', in *Religion and Rational Theology*, translated by Allen W. Wood and George Di Giovanni, Cambridge: Cambridge University Press, 1996, p. 260.
6 *Ibid.*, p. 259.
7 *Ibid.*, p. 255.
8 *Ibid.*, p. 240.
9 Fanny Defrance-Jublot, 'Le darwinisme au regard de l'orthodoxie catholique. Un manuscrit exhumé', *Revue d'histoire des sciences humaines*, no. 22, 2010, p. 233.
10 Benedict XVI, 'Faith, Reason and the University: Memories and Reflections', Aula Magna of the University of Regensburg, 12 September 2006; https://w2.vatican.va/content/benedict-xvi/en/speeches/2006/septem ber/documents/hf_ben-xvi_spe_20060912_university-regensburg.html.
11 Luca Bianchi, *Censure et liberté*, pp. 89–162.
12 Averroes, *L'Islam et la Raison*, translated by Marc Geoffroy, Paris: Flammarion, coll. GF Flammarion, 2000, p. 13.
13 Cited by Bianchi, *Censure et liberté*, p. 104.
14 Cited by Jean-Barthélémy Hauréau, 'Grégoire IX et la philosophie d'Aristote', *Comptes rendus des séances de l'Académie des Inscriptions et Belles-Lettres*, 16e année, 1872, pp. 531–2, our translation.
15 Maxime Dury, *La Censure*, Paris: Publisud, 1995.
16 See Alain Boureau, *Théologie, science et censure au XIIIe siècle*, Paris: Les Belles Lettres, 2008; and Bianchi, *Censure et liberté*; Edward Grant, *God and Reason in the Middle Ages*, Cambridge: Cambridge University Press, 2001, pp. 12–13, 214–17; Sara L. Uckelman, 'Logic and the Condemnation of 1277', *Journal of Philosophical Logic*, vol. 39, no. 2, 2010, pp. 201–27.
17 *La condamnation parisienne de 1277*, Latin text, translation, introduction and commentaries by D. Piché, Paris: Vrin, 1999, p. 73; our translation, emphasis added. Items in Table 1.1 come from Edward Grant (ed.), *A Source Book in Medieval Science*, Cambridge, MA: Harvard University Press, 1974, pp. 45–50; except for items 4, 40, 63, which we have translated in English using the French and Latin versions in *La condemnation parisienne*, pp. 81–101.
18 Cited by Bianchi, *Censure et liberté*, p. 84, our translation; see also William

J. Courtenay, 'Inquiry and Inquisition: Academic Freedom in Medieval Universities', *Church History*, vol. 58, no. 2, 1989, pp. 168–81.

19 Peter Godman, *Histoire secrète de l'Inquisition. De Paul III à Jean-Paul II*, translated by Cécile Deniard, Paris: Perrin, coll. Tempus, 2008, pp. 330–1.

20 Christopher Upham, 'The Influence of Aquinas,' in Brian Davies and Eleonore Stump (eds), *Oxford Handbook to Aquinas*, Oxford: Oxford University Press, 2012, p. 518.

21 Cited by Pierre Thibault, *Savoir et Pouvoir. Philosophie thomiste et politique cléricale au XIXe siècle*, Sainte-Foy: Presses de l'Université Laval, 1972, p. 143.

22 Francesco Beretta, 'Une deuxième abjuration de Galilée ou l'inaltérable hiérarchie des disciplines', *Bruniana et Campanelliana*, no. 9, 2003, pp. 15–16, our translation.

23 Owen Gingerich, *The Book Nobody Read. Chasing the Revolutions of Nicolaus Copernicus*, New York: Walker and Company, 2004.

24 Astronomy was then a branch of mathematics and the ancient Greek term *mathematicos* referred to both astronomy and mathematics.

25 Nicolaus Copernicus, *On the Revolutions of the Heavenly Spheres*, translated by Charles Glenn Wallis, Great Books of the Western World, London: Encyclopedia Britannica, 1952, pp. 506–9.

26 Owen Gingerich, *The Eye of Heaven: Ptolemy, Copernicus, Kepler*, New York: AIP, 1993, p. 167.

27 For more details, see Geoffrey Blumenthal, 'Copernicus's Publication Strategy in the Contexts of Imperial and Papal Censorship and of Warmian Diplomatic Precedents', *Science in Context*, vol. 29, no. 2, June 2016, pp. 151–78.

28 Massimo Bucciantini, *Galilée et Kepler. Philosophie, cosmologie et théologie à l'époque de la Contre-Réforme*, Paris: Les Belles Lettres, 2008, p. 99.

29 Reijer Hooykaas, *G.J. Rheticus' Treatise on Holy Scripture and the Motion of the Earth*, Amsterdam: North Holland, 1984.

30 Georgii Joachimi, *Narratio Prima*, critical edition, translated in French with commentaries by Henri Hugonard-Roche and Jean-Pierre Verdet, with the collaboration of Michel-Pierre Lerner and Alain Seconds, Varvosie: Académie polonaise des sciences, 1982, p. 91; see also Alcinoos, *Enseignement des doctrines de Platon*, Paris: Les Belles Lettres, 1990, p. 1.

31 Thomas Campanella, *The Defense of Galileo*, translated by Grant McColley, New York: Richwood Publishing, 1976.

32 Michel-Pierre Lerner, 'Aux Origines de la polémique anticopernicienne (I)', *Revue des sciences philosophiques et théologiques*, vol. 86, 2002, p. 683.

33 Copernicus, *On the Revolutions of the Heavenly Spheres*, p. 505, translation slightly simplified.

34 Nick Jardine, *The Birth of the History and Philosophy of Science*, Cambridge: Cambridge University Press, 1984.

35 F. Jamil Ragep, 'Freeing Astronomy from Philosophy: An Aspect of Islamic Influence on Science', *Osiris*, vol. 16, 2001, pp. 49–64.

36 For a detailed analysis, see Pierre-Michel Lerner, 'Copernic suspendu et corrigé. Sur deux décrets de la Congrégation romaine de l'index (1616–1620)', *Galilaena*, vol. 1, 2004, pp. 21–89.

37 Johannes Kepler, *Mysterium Cosmographicum. The Secret of the Universe*, translated by A.M. Duncan, New York: Abaris Books, 1981, p. 75.

38 Edward Rosen, 'Kepler and the Lutheran Attitude Towards Copernicanism

in the Context of the Struggle Between Science and Religion', *Vistas in Astronomy*, vol. 18, 1975, pp. 317–38; see Kepler's letter to Maestlin in June of 1598, cited on p. 329.

39 Copernicus, *On the Revolutions of the Heavenly Spheres*, p. 506.
40 Jean Kepler, *Selections from Kepler's Astronomia Nova*, selected, translated and annotated by William H. Donahue, Santa Fe: Green Lion Press, 2004, p. 25.
41 Rosen, 'Kepler and the Lutheran Attitude Towards Copernicanism', p. 326.
42 Valentine Zuber (ed.), *Michel Servet (1511–1553)*. *Hérésie et pluralisme du XVIe au XXIe siècle: actes du colloque de l'École pratique des hautes études, 11–13 décembre 2003*, Paris: Honoré Champion, 2007; Catherine Santschi, 'Les instances de contrôle protestantes', in Catherine Brice and Antonella Romano (eds), *Sciences et Religions. De Copernic à Galilée (1540–1610): actes du colloque international*, Rome: École française de Rome, 1999, pp. 467–71.
43 For details, see Stillman Drake, *Galileo at Work. His Scientific Biography*, Chicago, IL: University of Chicago Press, 1978; John L. Heilbron, *Galileo*, Oxford: Oxford University Press, 2010; Massimo Bucciantini, Michele Camerota and Franco Giudice, *Galileo's Telescope, A European Story*, translated by Catherine Bolton, Cambridge, MA: Harvard University Press, 2015.
44 Galileo's Letter to the Grand Duchess Christina, in Maurice A. Finocchiaro (ed.) *The Galileo Affair. A Documentary History*, Berkeley, CA: University of California Press, 1989, p. 89.
45 *Ibid.*, p. 90.
46 *Ibid.*, p. 96.
47 *Ibid.*, p. 101.
48 *Ibid.*, p. 95.
49 Augustine cited by Galileo in *ibid.*, p. 88, repeated later on p. 105.
50 *Ibid.*, pp. 90–1.
51 *Ibid.*, p. 99.
52 *Ibid.*, pp. 100–1.
53 Michel-Pierre Lerner, 'Vérité des philosophes et vérité des théologiens selon Tomasso Campanella o.p.', *Freiburger Zeitschrift für Philosophie und Theologie*, vol. 48, 2001, p. 297.
54 G.V. Coyne and U. Baldini, 'The Young Bellarmine's Thoughts on the World System', in C.V. Coyne, M. Heller and J. Zycinki (eds), *The Galileo Affair: A Meeting of Faith and Science*, Cité du Vatican: Specola Vaticana, 1985, pp. 103–9.
55 Godman, *Histoire secrète de l'Inquisition*, p. 92; Pietro Redondi, *Galileo Heretic*, Princeton, NJ: Princeton University Press, 1989, p. 6; Bernard Bourdin, *La Genèse théologico-politique de l'État moderne*, Paris: Presses universitaires de France, 2004.
56 Bellarmine to Foscarini, 12 April 1615, in Finocchiaro, *The Galileo Affair*, p. 67. For more details, see Richard J. Blackwell, *Galileo, Bellarmine and the Bible*, Notre Dame, IN: University of Notre Dame Press, 1991.
57 *Ibid.*, p. 67.
58 Galileo's Letter to the Grand Duchess, in Finocchiaro, *The Galileo Affair*, pp. 95–6.
59 Bellarmine to Foscarini, 12 April 1615, in Finocchiaro, *The Galileo Affair*, p. 68.

60 On this question, see Beretta, 'Une deuxième abjuration de Galilée', p. 11.
61 Cesi to Galileo, 12 January 1615, *Le Opere di Galileo Galilei*, vol. 12, p. 129. Our translation.
62 Ciampoli to Galileo, 28 February 1615, in *Le Opere di Galileo Galilei*, vol. 12, p. 146.
63 Pierre Duhem, *To Save the Phenomena. An Essay on the Idea of Physical Theory from Plato to Galileo*, translated by Edmund Dolan and Chaninah Maschler; with an Introduction by Stanley L. Jaki, Chicago, IL: University of Chicago University Press, 2015.
64 Elizabeth L. Eisenstein, *The Printing Press as an Agent of Change*, Cambridge: Cambridge University Press, 1976.
65 Bernard Gui, *The Inquisitor's Guide: A Medieval Manual on Heretics*, edited by Janet Shirley, Welwyn Garden City: Ravenhall Books, 2006.
66 Godman, *Histoire secrète de l'Inquisition*, p. 94; see also Karen Sullivan, *The Inner Lives of Medieval Inquisitors*, Chicago, IL: University of Chicago Press, 2011.
67 Nicolau Eymerich and Francisco Peña, *Le Manuel des inquisiteurs*, translated by Louis Sala-Molins, Paris: Albin Michel, 2001, pp. 76–7.
68 *Ibid.*, p. 77.
69 *Ibid.*, p. 92.
70 *Ibid.*, p. 96.
71 *Ibid.*, pp. 92–3.
72 Francesco Beretta, 'Le siège apostolique et l'affaire Galilée: relectures romaines d'une condamnation célèbre', *Roma moderna e contemporanea*, vol. 7, no. 3, 1999, p. 422; see also Francesco Beretta, 'Galileo, Urbain VIII, and the Prosecution of Natural Philosophers', in Ernan McMullin (ed.), *The Church and Galileo*, Notre Dame, IN: University of Notre Dame Press, 2007, pp. 234–61.
73 Lorini's complaint, 7 February 1615, in Finocchiaro, *The Galileo Affair*, p. 134.
74 Olaf Pedersen, *Galileo and the Council of Trent*, Rome: Vatican Observatory Publications, *Studi Galileiani*, vol. 1, no. 6, 1991, p. 8.
75 Galileo to Castelli, 21 December 1613, in Finocchiaro, *The Galileo Affair*, p. 49.
76 Consultant's Report on the Letter to Castelli (1615), in Finocchiaro, *The Galileo Affair*, p. 136.
77 Caccini's Deposition, 20 March 1615, in Finocchiaro, *The Galileo Affair*, p. 137.
78 *Ibid.*, pp. 141–6.
79 Consultant's Report on Copernicanism, 24 February 1616, in Finocchiaro, *The Galileo Affair*, p. 146.
80 Inquisition Minutes, 25 February 1616, in Finocchiaro, *The Galileo Affair*, p. 147.
81 *Ibid.*, p. 148.
82 Decree of the Index, 5 March 1616, in Finocchiaro, *The Galileo Affair*, p. 148.
83 *Ibid.*, p. 149.
84 Drake, *Galileo at Work*, p. 256.
85 Decree of the Index, p. 148.
86 *Ibid.*, p. 149.

87 On the censured copies of Copernicus' book, see Gingerich, *The Book Nobody Read.*

88 Bucciantini, *Galilée et Kepler*, p. 358. For a list of permissions to read books prohibited by the Index, see Ugo Baldini and Leen Spruit (eds) *Catholic Church and Modern Science: Documents from the Archives of the Roman Congregations of the Holy Office and the Index*, vol. 1, t. 1, Sixteenth-century documents, Roma: Libreria Editrice Vaticana, 2009.

89 Nick Jardine, *The Birth of The History and Philosophy of Science.*

90 Dini to Galileo, 7 March 1616, in Finocchiaro, *The Galileo Affair*, p. 58.

91 See the list of corrections in Finocchiaro, *The Galileo Affair*, pp. 200–2. For a detailed analysis of them, see Lerner, 'Copernic suspendu et corrigé'.

92 Galileo to the Tuscan Secretary of State, 6 March 1616, in Finocchiaro, *The Galileo Affair*, p. 151.

93 Cardinal Bellarmine's Certificate, 26 May 1616, in Finocchiaro, *The Galileo Affair*, p. 153.

94 Cited by Bucciantini, *Galilée et Kepler*, p. 325, our translation.

95 Annibale Fantoli, *Galileo. For Copernicus and for the Church*, second edition, Rome: Vatican Observatory Publications, *Studi Galileiani*, vol. 3, 1996, pp. 323–7.

96 Johannes Kepler, 'Repontio ad ingoli disputaionem ... inédit 1618', in Pierre-Noël Mayaud, *Le Conflit entre l'astronomie nouvelle et l'Écriture Sainte aux XVIᵉ et XVIIᵉ siècles*, vol. 3, Paris: Honoré Champion, 2005, p. 263, our translation.

97 *Ibid.*, pp. 263–4.

98 Johannes Kepler, 'Adminitio ad bibliopolas exteros, praesetrim Italos, de Opere Harmonico', in Mayaud, *Le Conflit entre l'astronomie nouvelle et l'Écriture Sainte*, vol. 3, pp. 265–6, our translation.

99 Cited in Lerner, 'Vérité des philosophes et vérité des théologiens', p. 298.

100 Maurice A. Finocchiaro, *Defending Copernicus and Galileo. Critical Reasoning in the Two Affairs*, Dordrecht: Springer, 2010, pp. 72–6; Fantoli, *For Copernicanism and for the Church*, pp. 264–5; see the document in Finocchiaro, *The Galileo Affair*, pp. 200–2.

101 Galileo to Cesi, 9 October 1623, *Le opere di Galileo Galilei*, vol. 13, p. 135, our translation.

102 Fantoli, *For Copernicanism and for the Church*, p. 286.

103 Drake, *Galileo at Work*, p. 287.

104 See his 'Discourse on the Tides', in Finocchiaro, *The Galileo Affair*, pp. 119–33.

105 Drake, *Galileo at Work*, p. 289.

106 *Ibid.*, p. 291.

107 Kenneth J. Howell, 'The Role of Biblical Interpretation in the Cosmology of Tycho Brahe', *Studies in History and Philosophy of Science*, vol. 29, 1998, pp. 515–37.

108 Fantoli, *For Copernicanism and for the Church*, pp. 327–8.

109 Drake, *Galileo at Work*, p. 300.

110 Francesco Beretta, '"Omnibus Christianae, Catholicaeque Philosophiae amantibus. D.D." Le Tractacus syllepticus de Melchior Inchofer, censeur de Galilée', *Freiburger Zeitschrift für Philosophie und Theologie*, vol. 48, 2001, pp. 308–9; see also Richard J. Blackwell, *Behind the Scenes at Galileo's Trial*, Notre Dame, IN: University of Notre Dame Press, 2006.

111 Galileo to Cesi, 8 June 1624, *Le opere di Galileo Galilei*, vol. 13, p. 182, our translation.
112 Drake, *Galileo at Work*, p. 291.
113 Galileo's Reply to Ingoli, in Finocchiaro, *The Galileo Affair*, p. 155, emphasis added.
114 *Ibid.*, p. 156.
115 *Ibid.*, p. 197.
116 Guiducci to Galileo, 18 April 1625, in Finocchiaro, *The Galileo Affair*, p. 205.
117 For details, see Fantoli, *For Copernicanism and for the Church*, pp. 335–47; also Mario Biagioli, *Galileo Courtier*, Chicago, IL: University of Chicago Press, 1993.
118 Galileo to the Tuscan Secretary of State, 3 May 1631, in Finocchiaro, *The Galileo Affair*, p. 211.
119 The Italian original reads: 'proponendo *indeterminatamente* le ragioni Filosofiche e Naturali tanto per l'una, quanto per l'altra parte'; Galileo Galilei, *Dialogue Concerning the Two Chief World Systems*, second edition, translated by Stillman Drake, Berkeley, CA: University of California Press, 1967, p. 1. The translation is taken from Fantoli, *For Copernicanism and for the Church*, p. 344, emphasis added.
120 Galileo Galilei, *Dialogue Concerning the Two Chief World Systems*, p. 5, emphasis added.
121 *Ibid.*, p. 5.
122 *Ibid.*, p. 6.
123 *Ibid.*, p. 6.
124 *Ibid.*, p. 464.
125 Fantoli, *For Copernicanism and for the Church*, pp. 392–4.
126 I would like to thank Raymond Fredette, a Galileo specialist, for helping me to appreciate more fully this aspect of the question.
127 Fantoli, *For Copernicanism and for the Church*, p. 402.
128 *Ibid.*, p. 408.
129 *Ibid.*, p. 420.
130 *Ibid.*
131 Galileo's First Deposition, 12 April 1633, in Finocchiaro, *The Galileo Affair*, p. 262.
132 Fantoli, *For Copernicanism and for the Church*, p. 427.
133 Eymerich and Peña, *Le Manuel des inquisiteurs*, p. 82.
134 *Ibid.*, p. 126.
135 Fantoli, *For Copernicanism and for the Church*, p. 428.
136 Eymerich and Peña, *Le Manuel des inquisiteurs*, p. 82.
137 *Ibid.*, p. 278.
138 Galileo's Second Deposition, 30 April 1633, in Finocchiaro, *The Galileo Affair*, p. 278.
139 *Ibid.*, pp. 278–9.
140 Sentence, 22 June 1633, in Finocchiaro, *The Galileo Affair*, p. 291.
141 *Ibid.*, p. 292.
142 Emile Namer, *L'Affaire Galilée*, Paris: Gallimard/Juliard, coll. 'Archives', 1975, p. 230.
143 Fantoli, *For Copernicanism and for the Church*, p. 487.
144 Godman, *Histoire secrète de l'Inquisition*, p. 39.

145 Galileo to Diodati, 25 July 1634, *Le opere di Galileo Galilei*, vol. 16, p. 116, our translation.
146 Cited in Jules Speller, *Galileo's Inquisition Trial Revisited*, Frankfurt: Peter Lang, 2008, p. 355.
147 *Ibid.*, p. 355.
148 Léon Garzend, *L'Inquisition et l'Hérésie*, Paris: Desclée de Brower, 1912, p. 64.

2. Copernicus and Galileo: Thorns in the Sides of Popes

1 Peiresc to Cardinal Francesco Barberini, 31 January 1635, in Franco Lo Chiatto and Sergio Marconi, *Galilée entre le pouvoir et le savoir*, Aix-en-Provence, Alinéa, 1988, p. 245; our translation.
2 John Paul II, 'On the Centenary of the Birth of Albert Einstein', Discourse of His Holiness Pope John Paul II given on 10 November 1979 at the Plenary Academic Session to commemorate the centenary of the birth of Albert Einstein, https://www.ewtn.com/library/PAPALDOC/JP2ALEIN. HTM.
3 For a recent biography, see Peter N. Miller, *L'Europe de Peiresc. Savoir et vertu au XVIIᵉ siècle*, Paris: Albin Michel, 2015.
4 Diodati to Gassendi, 10 November 1634, in Antonio Favaro (ed.), *Le Opere di Galileo Galilei*, Florence: G. Barbera, vol. 16, 1905, p. 153. Speaking of Galileo, Diodati wrote: 'I will tell you nothing more concerning his sufferings other than what I wrote to Mister Peiresc that if, through the connexions he has with Monsignor Cardinal Barberini, he could intercede in his favour to obtain some moderation of these great hardships, and make him get what he was given hope for (namely the release of his restriction in his house and freedom to be able to move to Florence and elsewhere), he would thus make a contribution of great merit and memorable charity'.
5 The original letters as well as Barberini's response are in Favaro (ed.), *Le Opere di Galileo Galilei*, vol. 16, passim. The dates used here are those established by Favaro; our translation.
6 Galileo to Peiresc, 16 March 1635, in Chiatto and Marconi, *Galilée entre le pouvoir et le savoir*, p. 259; our translation. See also the letter from Peiresc to Gassendi, 26 May 1635, in Favaro, *Le Opere di Galileo Galilei*, vol. 16, p. 268.
7 Domenico Bertoloni Meli, 'Leibniz on the Censorship of the Copernican System', *Studia Leibnitiana*, vol. 20, 1988, p. 21.
8 Christoph von Rommel (ed.), *Leibniz und Landgraf Ernst von Hessen-Rheinfels. Ein ungedruckter Briefwechsel über religiöse und politische Gegenstände*, Frankfurt: Literarische Anstalt, 1847, pp. 200–2; our translation.
9 Leibniz, *New Essays on Human Understanding*, translated by Alfred Gideon Langley, London: The Macmillan Company, 1896, book four, chapter XX, p. 614.
10 Cited by Pierre-Noël Mayhaud, *La Condamnation des livres coperniciens et sa révocation à la lumière de documents inédits des Congrégations de l'Index et de l'Inquisition*, Rome: Université pontificale grégorienne, 1997, p. 176; our translation.

11 *Ibid.*, p. 175.
12 Catherine Maire, 'L'entrée des "Lumières" à l'Index: le tournant de la double censure de *l'Encyclopédie* en 1759', *Recherches sur Diderot et sur l'Encyclopédie*, no. 42, 2007, pp. 108–39.
13 Peter Godman, *Histoire secrète de l'Inquisition. De Paul III à Jean-Paul II*, translated by Cécile Deniard, Paris: Perrin, coll. Tempus, 2008, pp. 254–5.
14 Cited in Mayaud, *La Condamnation des livres coperniciens*, p. 121; our translation.
15 For a detailed analysis of the corrections made to the *Revolutionibus*, see Pierre-Michel Lerner, 'Copernic suspendu et corrigé. Sur deux décrets de la Congrégation Romaine de l'Index (1616–1620)', *Galilaena*, vol. 1, 2004, pp. 21–89.
16 Cited in Mayaud, *La Condamnation des livres coperniciens*, p. 121, our translation.
17 Jean-Marc Ginoux and Christian Gerini, 'Poincaré et la rotation de la Terre', *Pour la science*, no. 417, July 2012, pp. 2–5; Jacques Gapaillard, *Et pourtant elle tourne. Le mouvement de la Terre*, Paris: Seuil, 1993.
18 Cited in Annibale Fantoli, *Galileo. For Copernicus and for the Church*, second edition, Rome: Vatican Observatory Publications, *Studi Galileiani*, vol. 3, 1996, pp. 491–2.
19 Michael Segre, 'The Never-Ending Galileo Story', in Peter Machamer (ed.), *The Cambridge Companion to Galileo*, Cambridge: Cambridge University Press, 1998, p. 392.
20 Fantoli, *For Copernicus and for the Church*, p. 493; Antonio Favaro (ed.), *Le Opere di Galileo Galilei*, Florence: G. Barbera, vol. 19, 1907, p. 399.
21 Paolo Galluzzi, 'The Sepulchers of Galileo: The "Living" Remains of a Hero of Science', in Machamer (ed.), *The Cambridge Companion to Galileo*, pp. 433–5. In their book *Galileo in Rome* (Oxford: Oxford University Press, 2003), William R. Shea and Mariano Artigas, write, on the contrary, that the ceremony was carried out 'in the presence of ecclesiastical authorities' (p. 200).
22 Mayaud, *La Condamnation des livres coperniciens*, p. 179.
23 Maria Pia Donato, 'Les doutes de l'inquisiteur: philosophie naturelle, censure et théologie à l'époque moderne', *Annales, Histoire, Sciences sociales*, vol. 64, 2009/1, p. 40.
24 Mayaud, *La Condamnation des livres coperniciens*, p. 215.
25 Many examples of Copernican books published are given in *ibid.*, pp. 218–33.
26 Cited in Fantoli, *For Copernicus and for the Church*, p. 498.
27 Cited in Fantoli, *Galilée. Pour Copernic et pour l'Église*, Rome: Publications de l'Observatoire du Vatican, *Studi Galileiani*, vol. 5, 2001, p. 343; our translation. This quotation is absent from the English version of the book published in 1996.
28 Owen Chadwick, *Catholicism and History: The Opening of the Vatican Archives*, Cambridge: Cambridge University Press, 1978, pp. 14–15.
29 *Ibid.*, pp. 20–1; Francesco Beretta, 'Le siège apostolique et l'affaire Galilée: relectures romaines d'une condamnation célèbre', *Roma moderna e contemporanea*, vol. 7, no. 3, 1999, p. 443; see also Francesco Beretta, 'Le procès de Galilée et les archives du Saint-Office. Aspects judiciaires et théologiques d'une condamnation célèbre', *Revue des sciences philosophiques et théologiques*, vol. 83, 1999, p. 467 and 'The Documents of Galileo's Trial: Recent Hypothesis and Historical Criticism', in Ernan McMullin (ed.), *The*

Church and Galileo, Notre Dame, IN: University of Notre Dame Press, 2007, pp. 191–212. Part of the correspondence between Paris and Rome can be found in L. Sandret, 'Le manuscrit original du procès de Galilée', *Revue des questions historiques*, vol. 22, 1877, pp. 551–9.

30 Fantoli, *For Copernicus and for the Church*, p. 501.
31 *Ibid.*, pp. 501–2.
32 For a detailed analysis, see Maurice A. Finocchiaro, *Retrying Galileo, 1633–1992*, Berkeley, CA: University of California Press, 2005.
33 Henri de l'Épinois, *Les Pièces du procès de Galilée précédées d'un avant-propos*, Paris: Société générale de librairie catholique, 1877, p. v. This is a revised and augmented edition of the first publication of these documents in the July 1867 issue of *Revue des questions historiques*.
34 Finocchiaro, *Retrying Galileo*, p. 262.
35 Antonio Favaro (ed.), *Le Opere di Galileo Galilei*, Florence: G. Barbera, 1890–1909. This edition is freely accessible on the internet site of the French Bibliothèque nationale at gallica.bnf.fr.
36 Pnina G. Abir-Am and Clark A. Elliot (ed.), *Commemorative Practices in Science: Historical Perspectives on the Politics of Collective Memory*, Osiris, vol. 14, Chicago, IL: University of Chicago Press, 1999.
37 Fantoli, *For Copernicus and for the Church*, p. 523.
38 Cited in *ibid.*, p. 504.
39 For more details on this affair, see Finocchiaro, *Retrying Galileo*, pp. 318–26.
40 Cited in *ibid.*, p. 324.
41 On Maccarrone's career, see Thomas F.X. Noble, 'Michele Maccarrone on the Medieval Papacy', *The Catholic Historical Review*, vol. 80, 1994, pp. 518–33.
42 Finocchiaro, *Retrying Galileo*, pp. 330–7.
43 Fantoli, *For Copernicus and for the Church*, pp. 526, 529; Finocchiaro, *Retrying Galileo*, p. 329.
44 Jean-Paul Messina, *Evêques africains au concile Vatican II, 1959–1965. Le cas du Cameroun*, Paris: Karthala, 2000, p. 138.
45 Léon-Arthur Elchinger, *L'Âme de l'Alsace et son avenir*, Strasbourg: La Nuée bleue, 1992, p. 167; our translation.
46 Cited in Henri Fesquet, 'Vatican II et la culture', *Le Monde*, 6 November 1964; our translation.
47 Cited in Elchinger, *L'Âme de l'Alsace et son avenir*, p. 167, our translation.
48 Cited in Alberto Melloni, 'Galileo al Vaticao II', in Massimo Bucciantini, Michele Camerota and Franco Guidice (eds), *Il Caso Galileo. Una rilettura storica, filosofica, teologica*, Florence: Leo S. Olschki, 2011, p. 482; our translation.
49 *Ibid.*, pp. 482–3.
50 For a more detailed analysis of these debates, see Melloni, 'Galileo al Vaticano II', pp. 461–90.
51 Cited in Finocchiaro, *Retrying Galileo*, p. 330.
52 Georges Gusdorf, *La Révolution galiléenne*, vol. 1, Paris: Payot, 1969, p. 133.
53 John Paul II, 'On the Centenary of the Birth of Albert Einstein', Discourse of His Holiness Pope John Paul II given on 10 November 1979 at the Plenary Academic Session to commemorate the centenary of the birth of Albert Einstein, https://www.ewtn.com/library/PAPALDOC/JP2ALEIN.HTM.

54 *Ibid.*
55 George V. Coyne, S.J., 'Galileo Judged. Urbain VIII to John Paul II', in Bucciantini, Camerota and Guidice (eds), *Il Caso Galileo*, p. 493.
56 Paul Poupard, 'Compte rendu des travaux de la commission pontificale d'études de la controverse ptolémo-copernicienne aux XVIᵉ–XVIIᵉ siècles', in Paul Poupard (ed.), *Après Galilée. Science et foi: nouveau dialogue*, Paris: Desclée de Brower, 1994, pp. 93–7.
57 Finocchiaro, *Retrying Galileo*, pp. 353–7; for a detailed critique, see Annibale Fantoli, 'Galileo and the Catholic Church: A Critique of the "Closure" of the Galileo Commission's Work', *Studi Galileiani*, vol. 4, no. 1, 2002.
58 Coyne, 'Galileo Judged', p. 498.
59 *Ibid.*, pp. 493–4.
60 John Paul II, 'Discourse to the Pontifical Academy of Science, 31 October 1992', http://bertie.ccsu.edu/naturesci/Cosmology/GalileoPope.html.
61 Fantoli, *For Copernicus and for the Church*, p. 511; see also Fantoli, *The Case of Galileo. A Closed Question?*, Notre Dame, IN: University of Notre Dame Press, 2012.
62 John Paul II, 'Discourse to the Pontifical Academy of Science', 31 October 1992.

3. God: From the Centre to the Periphery of Science

1 Émile Durkheim, *The Division of Labor in Society*, translated by W.D. Halls, New York: The Free Press, 1984, p. 119.
2 Yves Gingras, Peter Keating and Camille Limoges, *Du scribe au savant. Les porteurs du savoir de l'Antiquité à la révolution industrielle*, Montréal: Boréal, 1998, p. 265.
3 Frank E. Manual, *A Portrait of Newton*, Washington, DC: New Republic Books, 1979, pp. 119–20.
4 Éric Brian and Christiane Demeulenaere-Douyère (eds), *Règlement, usages et science dans la France de l'Absolutisme*, Paris: Technique & Documentation, 2002.
5 Cited in Fanny Defrance-Jublot, 'Question laïque et légitimité scientifique en préhistoire. La revue *L'Anthropologie* (1890–1910)', *Vingtième siècle. Revue d'histoire*, no. 87, 2005/3, p. 82.
6 On the social norms of science, see Robert K. Merton, *The Sociology of Science*, Chicago, IL: University of Chicago Press, 1973.
7 Trevor McClaughlin, 'Le concept de sciences chez Jacques Rohault', *Revue d'histoire des sciences*, vol. 30, no. 3, 1977, pp. 225–40.
8 Jacques Rohault, *Entretiens sur la philosophie*, Paris: Michel Le Petit, 1671, pp. 12–13; our translation.
9 *Ibid.*, p. 10.
10 Jack Morrell and Arnold Thackray, *Gentlemen of Science: Early Years of the British Association for the Advancement of Science*, Oxford: Oxford University Press, 1981, pp. 224–45.
11 John Henry Newman, 'The Usurpations of Reason', sermon preached 11 December 1831, reprinted in *Fifteen Sermons Preached Before the University of Oxford Between A.D. 1826 and 1843*, Third edition, London: Rivingtons, 1872, p. 72.

12 Mordechai Feingold (ed.), *Jesuit Science and the Republic of Letters*, Cambridge, MA: MIT Press, 2003.

13 John Henry Newman, 'Faith and Reason, Contrasted as Habits of Mind', Epiphany sermon, 1839, in *Fifteen Sermons*, p. 194.

14 See, for example, T.F. Torrance, 'Christian Faith and Physical Science in the Thoughts of James Clerk Maxwell', in T.F. Torrance (ed.), *Transformation and Convergence in the Frame of Knowledge: Explorations in the Interrelations of Scientific and Theological Enterprise*, Grand Rapids, MI: Eerdmans, 1984, pp. 215–42.

15 We will return to this issue in chapter 5.

16 Hans Reichenbach, *Experience and Prediction: An Analysis of the Foundations and the Structure of Knowledge*, Chicago, IL: University of Chicago Press, 1938. For a series of studies on this now classical distinction, see Jutta Schickore and Friedrich Steinle (ed.), *Revisiting Discovery and Justification: Historical and Philosophical Perspectives on the Context Distinction*, Dordrecht: Springer, 2006.

17 Charlotte Methuen, 'The Teachers of Johannes Kepler: Theological Impulses to the Study of the Heavens', in *Sciences et Religions. De Copernic à Galilée (1540–1610)*, collection de l'École française de Rome, 260, 1996, pp. 183–203.

18 Matthew Stanley, 'By Design: James Clerk Maxwell and the Evangelical Unification of Science', *British Journal for the History of Science*, vol. 45, March 2012, pp. 57–73.

19 Matthew Stanley, *Huxley's Church and Maxwell's Demon. From Theistic Science to Naturalistic Science*, Chicago, IL: University of Chicago Press, 2015, p. 2.

20 Pierre Gibert, *L'Invention critique de la Bible, XVᵉ –XVIIIᵉ siècle*, Paris: Gallimard, 2010.

21 Galileo Galilei, *Dialogue Concerning the Two Chief World Systems*, translated by Stillman Drake, Second Edition, Berkeley, CA: University of California Press, 1967, p. 422.

22 *Ibid.*

23 *Ibid.*, p. 29.

24 Steven Shapin, 'Pump and Circumstance: Robert Boyle's Literary Technology', *Social Studies of Science*, vol. 14, 1984, pp. 481–520.

25 Margaret G. Cook, 'Divine Artifice and Natural Mechanism: Robert Boyle's Mechanical Philosophy of Nature', *Osiris*, vol. 16, 2001, pp. 133–50.

26 Based on a full text search in the *Philosophical Transactions* of the Royal Society of London from 1665 to 1900, the results by century are: 1665–99: 108; 1700–99: 139; 1800–99: 18; perusing the context in which the word 'God' appear makes clear it is not invoked to *explain* phenomena.

27 Henry Oldenburg, 'Epistle Dedicatory', *Philosophical Transactions*, vol. 1 (1665–6).

28 Plato, *Laws*, 966d–e, http://www.perseus.tufts.edu/hopper/text?doc=Perseu s%3Atext%3A1999.01.0166%3Abook%3D12%3Asection%3D966e.

29 Véronique Le Ru, *La Nature, miroir de Dieu. L'ordre de la nature reflète-t-il la perfection du créateur?*, Paris: Vuibert, 2010, pp. 79–89.

30 John J. Dahm, 'Science and Apologetics in the Early Boyle Lectures', *Church History*, vol. 39, June 1970, pp. 172–86.

31 Nigel Smith, 'The Charge of Atheism and the Language of Radical

Speculation, 1640–1660', in Michael Hunter and David Wooton (eds), *Atheism from the Reformation to the Enlightenment*, Oxford: Clarendon Press, 1992, pp. 131–58; Georges Minois, *Histoire de l'athéisme*, Paris: Fayard, 1998.

32 See the list of lecturers and titles since 1692 on Wikipedia: https://en.wikipedia.org/wiki/Boyle_Lectures.

33 William H. Austin, 'Newton on Science and Religion', *Journal of the History of Ideas*, vol. 31, 1970, p. 522.

34 Peter J. Bowler, 'Complexity in Practice', *British Journal for the History of Science*, vol. 44, no. 2, June 2011, p. 277.

35 Isaac Newton, *Principia*, vol. 2, *The System of the World*, Berkeley, CA: University of California Press, 1973, p. 546; Isaac Newton, *Principia*, translated by I.B. Cohen and Anne Whitman, Berkeley, CA: University of California Press, 1999, p. 943.

36 Stephen D. Snobelen, '"God of Gods, and Lord of Lords": The Theology of Isaac Newton's General Scholium to the *Principia*', *Osiris*, vol. 16, 2001, p. 197, footnote 115.

37 Frank E. Manual, *A Portrait of Newton*, pp. 130–1.

38 *Ibid.*, p. 126.

39 For a detailed analysis of the general context that brought Newton to add the General Scholium in his *Principia*, see Larry Stewart, 'Seeing Through the Scholium: Religion and Reading Newton in the Eighteenth Century', *History of Science*, vol. 34, 1996, pp. 123–65; Stephen D. Snobelen 'To Discourse of God: Isaac Newton's Heterodox Theology and His Natural Philosophy', in Paul B. Wood (ed.), *Science and Dissent in England, 1688–1945*, Aldershot: Ashgate, 2004, pp. 39–65.

40 Spinoza, *Ethics*, translated by W. H. White, revised by A. H. Stirling, Ware, Hertfordshire: Wordsworth Editions, 2001, p. 39; this work was first published in 1677, the year of his death.

41 Cited in Le Ru, *La Nature, miroir de Dieu*, p. 88; our translation.

42 Henry Newman, *Fifteen Sermons*, p. 70.

43 *Ibid.*, p. 194.

44 John Henry Newman, 'Christianity and the Physical Sciences', A Lecture in the School of Medicine, November 1855, in *The Idea of a University*, Third edition, London: Basil Montagu Pickering, 1873, pp. 454–5.

45 On this anecdote, see Roger Hahn, 'Laplace and the Vanishing Role of God in the Physical Universe', in Harry Woolf (ed.), *The Analytic Spirit*, Ithaca, NY: Cornell University Press, 1981, pp. 85–95.

46 Pierre-Simon de Laplace, *The System of the World*, translated by Rev. Henry H. Harte, vol. II, Dublin: University Press, 1830, pp. 331–2. On the reception of Laplace's ideas by American Christians, see Ronald L. Numbers, *Creation by Natural Law. Laplace's Nebular Hypothesis in American Thought*, Seattle, WA: University of Washington Press, 1977.

47 *Ibid.*, p. 333.

48 In his letter to the Pope, Copernicus wrote that 'it is not unknown that Lactantius, otherwise a distinguished writer but hardly a mathematician speaks in an utterly childish fashion concerning the shape of the Earth when he laughs at those who have affirmed that the Earth has the form of a globe'; Copernicus, *On the Revolutions of the Heavenly Spheres*, translated

by Charles Glenn Wallis, Great Books of the Western World, London: Encyclopedia Britannica, 1952, p. 509.

49 Charles Coulston Gillispie, *Genesis and Geology: A Study in the Relations of Scientific Thought, Natural Theology and Social Opinion in Great Britain, 1790–1850*, Cambridge, MA: Harvard University Press, 1969.

50 Charles Lyell, *Principles of Geology*, vol. 1, Fifth Edition, London: John Murray, 1837, p. 59.

51 *Ibid.*, p. 64.

52 Buffon, *Natural History*, translated by William Smellie, second edition, vol. 1, London: W. Strahan and T. Cadell, 1785, p. 63.

53 *Ibid.*, p. 114.

54 *Ibid.*, p. 127.

55 *Ibid.*, pp. 129–30.

56 *Ibid.*, p. 130.

57 *Ibid.*, pp. 127–9.

58 *Ibid.*, p. 132.

59 Lyell, *Principles of Geology*, p. 43.

60 *Ibid.*, p. 44.

61 Charles Darwin, *On the Origin of Species by Means of Natural Selection*, third edition, London: John Murray, 1861, p. 100.

62 *Ibid.*, p. 305.

63 Darwin to Lyell, 28 March 1859, in Darwin Correspondence Project; https://www.darwinproject.ac.uk/.

64 William Whewell, *Astronomy and General Physics Considered with Reference to Natural Theology*, London: William Pickering, 1833, pp. 356–7.

65 Cited by Darwin in epigraph to his book *On the Origin of Species*. The case of Bacon's own censure first by the Roman Index in 1668 of his *Die dignitate et augmentis scientiarum*, 'until corrected' (*donec corrigatur*) and later in the hand of Joseph de Maistre (1753–1821), as a father of modern atheism for having separated out the study of nature from the study of God, is a further example of the continued resistance to the autonomy of science in the nineteenth century; see Joseph de Maistre, 'The Union of Religion and Science', in *An Examination of the Philosophy of Bacon: Wherein Different Questions of Rational Philosophy Are Treated*, translated and edited by Richard A. Lebrun, Montreal: McGill-Queen's University Press, 1998, pp. 270–91.

66 Kenneth J. Howell, *God's Two Books. Copernican Cosmology and Biblical Interpretation in Early Modern Science*, Notre Dame, IN: University of Notre Dame Press, 2002.

67 Darwin, *On the Origin of Species*, p. 524.

68 *Ibid.*, p. 525.

69 *Ibid.*, p. 514.

70 See the letter of Charles Kingsley to Darwin, 18 November 1859, in *The Correspondence of Charles Darwin*, vol. 7, 1858–1859, Cambridge: Cambridge University Press, 1992, pp. 379–80.

71 *Ibid.*, p. 515.

72 Francis Darwin (ed.), *The Life and Letters of Charles Darwin. Including an Autobiographical Chapter*, vol. 1, New York: D. Appleton & Co., 1887, p. 278.

73 Nora Barlow (ed.), *The Autobiography of Charles Darwin 1809–1882. With the Original Omissions Restored. Edited and with Appendix and Notes by his Grand-daughter Nora Barlow*. London: Collins, 1958, p. 87. Now accessible on line: http://darwin-online.org.uk/content/frameset?viewt ype=text&itemID=F1497&pageseq=1

74 *Ibid.*, p. 279.

75 Darwin, *On the Origin of Species*, p. 516.

76 Darwin, *Life and Letters*, pp. 75–6.

77 Jean-Baptiste Lamarck, *Zoological Philosophy*, translated by Hugh Elliot, London: Macmillan and Co, 1914, p. 170.

78 *Ibid.*, p. 170.

79 *Ibid.*, p. 171.

80 *Ibid.*, p. 173.

81 Jacques Arnould, *L'Église et l'histoire de la nature*, Paris: Cerf, 2000, p. 57.

82 Pietro Corsi, 'Idola Tribus: Lamarck, Politics and Religion in the Early Nineteenth Century', in Aldo Fasolo (ed.), *The Theory of Evolution and Its Impact*, Dordrecht: Springer-Verlag, 2012, p. 33.

83 Darwin, *On the Origin of Species*, p. xiii.

84 Boucher de Perthes, *Antiquités celtiques et antédiluviennes. Tome troisième*, Paris: Dumoulin, 1864, p. 112; for more detail, see Marc Groenen, *Pour une histoire de la préhistoire*, Paris: Jérôme Million, 1994, pp. 52–72.

85 Charles Lyell, *The Geological Evidences of the Antiquity of Man*, Fourth edition revised, London: John Murray, 1873, p. 552.

86 See François Laplanche, *La Bible en France entre mythe et critique, XVIe–XIXe siècle*, Paris: Albin Michel, 1994; see also Laplanche, *La Crise de l'origine. La science catholique des Évangiles et l'histoire au XXe siècle*, Paris: Albin Michel, 2006.

87 Laplanche, *La Bible en France*, pp. 141–2.

88 Philip Harwood, *German Anti-Supernaturalism. Six lectures on Strauss's 'Life of Jesus', Delivered at the Chapel in South Place, Finsbury*, London: Charles Fox, 1841.

89 Perrine Simon-Nahum, 'Le scandale de la *Vie de Jésus* de Renan. Du succès littéraire comme mode d'échec de la science', *Mil neuf cent*, no. 25, 2007, pp. 61–74.

90 Ernest Renan, *Studies of Religious History and Criticism*, translated by O. B. Frothingham, New York: Carleton Publishers, 1864, p. 171.

91 *Ibid.*, p. 42.

92 *Ibid.*, p. 45.

93 *Ibid.*, p. 42.

94 *Ibid.*, p. 54.

95 Charles Darwin, *The Descent of Man, and Selection in Relation to Sex*, New York: D. Appleton & Co., 1871, pp. 2–3.

96 *Ibid.*, p. 4.

97 *Ibid.*, p. 63.

98 *Ibid.*, p. 63.

99 *Ibid.*, p. 101.

100 *Ibid.*, pp. 101–2.

4. Science Censored

1 Paul Henri Thiry, Baron d'Holbach, *Théologie portative ou Dictionnaire abrégé de la religion chrétienne*, Paris: Rivages Poche, 2015, p. 93; our translation.
2 Descartes to Mersenne, late November 1633, in *Œuvres de Descartes*, edition Adam and Tannery, vol. 1, Paris: Vrin, 1996, pp. 270–1; translated into English by Jonathan Bennett, *Selected Correspondence of Descartes*, p. 27; http://www.earlymoderntexts.com/assets/pdfs/descartes1619_1.pdf.
3 Descartes to Mersenne, April 1633, translated into English by Jonathan Bennett, *Selected Correspondence of Descartes*, p. 29.
4 Descartes to Mersenne, December 1640, cited in Léon Petit, 'L'affaire Galilée vue par Descartes et Pascal', *Dix-septième siècle*, no. 28, 1955, p. 234; translated into English by Jonathan Bennett, *Selected Correspondence of Descartes*, p. 117.
5 Adrien Baillet, *Vie de Monsieur Descartes*, vol. 1, Paris: Daniel Hortemels, 1691, pp. 253–4; our translation.
6 Descartes to Mersenne, June or July 1635, translated into English by Jonathan Bennett, *Selected Correspondence of Descartes*, p. 33.
7 Jean-Robert Armogathe and Vincent Carraud, 'La première condamnation des œuvres de Descartes, d'après des documents inédits aux archives du Saint-Office', *Nouvelles de la République des Lettres*, no. 2, 2001, pp. 103–23.
8 Boulliau to Mersenne, 16 December 1644, http://users.clas.ufl.edu/ufhatch/pages/11-ResearchProjects/boulliau/06rp-b-lttrs.htm; see also Michel-Pierre Lerner, 'La réception de la condamnation de Galilée en France au XVIIᵉ siècle', in Jose Montesinos and Carlos Solis (eds), *Largo campo di filosofare. Eurosymposium Galileo 2001*, La Orotava: Fundación Canaria Orotava de Historia de la Ciencia, 2001, p. 532.
9 Isabelle Pantin, 'Premières répercussions de l'affaire Galilée en France chez les philosophes et les libertins', in Massimo Bucciantini, Michele Camerota and Franco Guidice (eds), *Il Caso Galileo. Una rilettura storica, filosofica, teologica*, Florence: Leo S. Olschki, 2011, p. 243.
10 Lerner, 'La réception de la condamnation de Galilée en France', p. 533; see also Lisa T. Sarasohn, 'French Reaction to the Condemnation of Galileo, 1632–1642', *The Catholic Historical Review*, vol. 74, 1988, pp. 34–54; Jane T. Tolbert, 'Peiresc and Censorship: The Inquisition and the New Science, 1610–1637', *The Catholic Historical Review*, vol. 89, 2003, pp. 24–38.
11 Mersenne to M. Rebours, November 1633, cited in B. Rochot, 'Remarques sur l'affaire Galilée', *Dix-septième siècle*, no. 30, 1956, p. 135; our translation.
12 Cited in Lerner, 'La réception de la condamnation de Galilée en France', p. 538; our translation.
13 *Ibid.*, p. 532.
14 Peiresc to Gassendi, 15 January 1634, in Antonio Favaro (ed.), *Le Opere di Galileo Galilei*, Florence: G. Barbera, vol. 16, 1905, pp. 14–15; our translation.
15 Mersenne to Peiresc, 28 July 1634, in *Le Opere di Galileo Galilei*, vol. 16, p. 119; our translation.

16 Jesús Martínez de Bujanda, Francis M. Higman and James K. Farge (eds), *Index de l'Université de Paris: 1544, 1545, 1547, 1549, 1551, 1556*, Geneva: Libraire Droz, 1985.

17 Cited in *ibid.*, p. 37; our translation.

18 Cited in Maurice A. Finocchiaro, *Retrying Galileo, 1633–1992*, Berkeley, CA: University of California Press, 2005, pp. 83–4.

19 On this episode, see Ivana Gambaro, *Astronomia e tecniche di ricerca nelle lettere di G.B. Riccioli ad A. Kircher*, Genova: CNR – Centro di Studio sulla Storia della Tecnica, 1989, pp. 39–42.

20 Cited in Armand Stévart, *Procès de Martin Étienne van Velden, professeur à l'université de Louvain*, Brussels: C. Muquardt, 1871, p. 60; our translation.

21 Jesús Martínez de Bujanda (ed.), *Index de l'Université de Louvain: 1546, 1550, 1558*, Geneva: Libraire Droz, 1986, p. 14.

22 Cited in George Monchamp, *Galilée et la Belgique*, Brussels: G. Moreau-Schouberechts, 1892, p. 216; our translation.

23 Christiaan Huygens, *Œuvres complètes*, vol. X, *Correspondance 1691–1695*, The Hague: Martinus Nijhoff, 1905, pp. 113–14, our translation. Also at: www.dbnl.org/tekst/huyg003oeuv10_01/huyg003oeuv10_01_0035.php.

24 Monchamp, *Galilée et la Belgique*, p. 311.

25 *Ibid.*, p. 314; our translation.

26 *Ibid.*, p. 315; our translation.

27 Stévart, *Procès de Martin Étienne van Velden*, p. 62; Monchamp, *Galilée et la Belgique*, p. 322.

28 Cited in Monchamp, *Galilée et la Belgique*, pp. 302–3; our translation.

29 Stephen Greenblatt, *The Swerve*, New York: W.W. Norton and Company, 2009.

30 Cited in Maurice A. Finocchiaro, *The Galileo Affair. A Documentary History*, Berkeley, CA: University of California Press, 1989, p. 203.

31 *Ibid.*, p. 204.

32 Cited in Pietro Redondi, *Galileo Heretic*, translated by Raymon Rosenthal, Princeton, NJ: Princeton University Press, 1987, p. 336.

33 Historian Francesco Beretta suggests that the failure of this denunciation is due to the fact that atomistic theses were not very well known in Italy at the time. See Beretta, 'Doctrine des philosophes, doctrine des théologiens et Inquisition au 17ᵉ siècle: aristétotélisme, héliocentrisme, atomisme', in *Vera Doctrina: zu Begrisgeschichte der Lehre von Augustinus bis Descartes*, Wiesbaden: Otto Harrassowitz, 2006, pp. 173–97, also at: https://hal.archives-ouvertes.fr/halshs-00453269/document. The report of the consultor has been published in Thomas Cerbu, 'Melchior Inchofer, "un homme rusé"', in Montesinos and Solis (eds), *Largo campo di filosofare*, pp. 587–611.

34 Lynn Thorndike, 'Censorship by the Sorbonne of Science and Superstition in the First Half of the Seventeenth Century', *Journal of the History of Ideas*, vol. 16, no. 1, 1955, pp. 119–25.

35 Redondi, *Galileo Heretic*, p. 240.

36 Cited in Sylvain Matton, 'Note sur quelques critiques oubliées de l'atomisme: à propos de la transubstantiation', *Revue d'histoire des sciences*, vol. 55, no. 2, 2002, p. 289; our translation.

37 Sophie Roux, 'Descartes atomiste?', in Romano Gatto and Egidio Festa

(eds), *Atomismo e continuo nel XVII secolo*, Naples: Vivarium, 2000, pp. 211–74.

38 Jean-Robert Armogathe and Vincent Carraud, 'La première condamnation des Œuvres de Descartes d'après les documents inédits aux archives du Saint-Office', *Nouvelles de la république des lettres*, vol. 2, 2001, pp. 103–23.

39 Jean-Robert Armogathe, 'Cartesian Physics and the Eucharist in the Documents of the Holy Office and the Roman Index (1671–6)', in Tad M. Schmaltz (ed.), *Receptions of Descartes: Cartesianism and Anti-Cartesianism in Early Modern Europe*, New York: Routledge, 2005, p. 147.

40 Nicola Borchi, 'La Métaphysique d'Antonio Genovesi face à la censure ecclésiastique de Rome', in Jacques Domenach (ed.), *Censure, autocensure et art d'écrire*, Brussels: Complexe, 2005, pp. 157–64.

41 Pius XII, *Humani Generis*, paragraph 26, English version on the website of the Vatican: http://w2.vatican.va/content/pius-xii/en/encyclicals/documents/hf_p-xii_enc_12081950_humani-generis.html.

42 Steven J. Dick, *Plurality of Worlds: The Origins of the Extraterrestrial Life Debate from Democritus to Kant*, Cambridge: Cambridge University Press, 1982, pp. 44–60.

43 Cited in Dick, *Plurality of Worlds*, p. 55.

44 Consultor report cited in Francesco Beretta, 'L'héliocentrisme à Rome à la fin du XVIIe siècle: une affaire d'étrangers?', in Antonella Romano (ed.), *Rome et la science moderne*, Rome: École française de Rome, 2008, p. 545; our translation.

45 John Wilkins, *A Discovery of a New World*, in 2 parts, London, 1684.

46 *Ibid.*, book 2, p. 3.

47 Ted Peters, 'The Implications of the Discovery of Extra-Terrestrial Life for Religion', *Philosophical Transactions: Mathematical, Physical and Engineering Sciences*, vol. 369, 13 February 2011, p. 644.

48 Pierre-Noël Mayhaud, *La Condamnation des livres coperniciens et sa révocation à la lumière de documents inédits des Congrégations de l'Index et de l'Inquisition*, Rome: Université pontificale grégorienne, 1997, p. 170; our translation.

49 J.M. de Bujanda, *Index librorum prohibitorum, 1600–1966*, Montreal: Médiaspaul, 2002, p. 498.

50 Jérôme de Lalande, *Astronomie des dames*, fourth edition, Paris: Ménard et Desenne fils, 1817, pp. 120–1; our translation.

51 Jacques Roger, *Buffon*, Paris: Fayard, 1989, pp. 250–1.

52 Buffon to l'abbé Le Blanc, 23 June 1750, in *Correspondance de Buffon*, electronic edition, letter L37, www.buffon.cnrs.fr; our translation.

53 Buffon, *Histoire naturelle*, vol. IV, Paris, 1753, p. v; our translation.

54 The complete list of problematic propositions is reproduced at the beginning of vol. IV of *Histoire naturelle*, pp. vii–ix; our translation.

55 Roger, *Buffon*, pp. 248–9.

56 *Histoire naturelle*, vol. IV, p. xii; our translation.

57 Buffon to de Brosses, 14 July 1760, *Correspondance de Buffon*, letter L73.

58 Buffon to l'abbé Le Blanc, 24 April 1751, *Correspondance de Buffon*, letter L41.

59 Buffon to Guéneau de Montbeillard, 15 November 1779, *Correspondance de Buffon*, letter L367; our translation.

60 Cited in Maria Pia Donato, 'Les doutes de l'inquisiteur: philosophie naturelle, censure et théologie à l'époque moderne', *Annales, Histoire, Sciences sociales*, vol. 64, 2009/1, p. 39; our translation.

61 De Bujanda, *Index librorum prohibitorum*, p. 742.

62 Peter Godman, *Histoire secrète de l'Inquisition. De Paul III à Jean-Paul II*, translated by Cécile Deniard, Paris: Perrin, coll. 'Tempus', 2008, p. 333.

63 De Bujanda, *Index librorum prohibitorum*, p. 498.

64 *Ibid.*, p. 176.

65 Norton Garfinkle, 'Science and Religion in England, 1790–1800: The Critical Response to the Work of Erasmus Darwin', *Journal of the History of Ideas*, vol. 16, no. 3, June 1955, 376–88.

66 Pietro Corsi, *Science and Religion: Baden Powell and the Anglican Debate, 1800–1860*, Cambridge: Cambridge University Press, 1988, p. 56; Peter G. Mudford, 'William Lawrence and The Natural History of Man', *Journal of the History of Ideas*, vol. 29, 1968, pp. 430–6.

67 The complete text is accessible online: http://www.papalencyclicals.net/ Pius09/p9syll.htm.

68 François Laplanche, *La Bible en France entre mythe et critique, XVIe–XIXe siècle*, Paris: Albin Michel, 1994, pp. 189–91.

69 François Lenormant, *The Beginnings of History According to the Bible and the Traditions of Oriental Peoples: From the Creation of Man to the Deluge*, Translated from the Second French edition by Francis Brown, New York: Charles Scribner and Sons, 1882, p. ix.

70 *Ibid.*, pp. ix–x.

71 *Ibid.*, p. x.

72 *Ibid.*, p. xi.

73 Laplanche, *La Bible en France*, p. 192.

74 *Providentissimus Deus*, encyclical of Pope Leo XIII on the Study of Holy Scripture, 18 November 1893; English text on the website of the Vatican: http://w2.vatican.va/content/leo-xiii/en/encyclicals/documents/hf_l-xiii_ enc_18111893_providentissimus-deus.html.

75 Pius X, July 3 1907, Lamentabili Sane, http://www.papalencyclicals.net/ Pius10/p10lamen.htm.

76 Harvey Hill, *The Politics of Modernism: Alfred Loisy and the Scientific Study of Religion*, Washington, DC: Catholic University of America Press, 2002.

77 Cited in François Laplanche, *La Crise de l'origine. La science catholique des Évangiles et l'histoire au XXe siècle*, Paris: Albin Michel, 2006, p. 12; our translation.

78 In Alfred Loisy, *Quelques lettres sur des questions actuelles et sur des événements récents*, près Montier-en-Der, Haute-Marne: Chez l'auteur, 1908, pp. 290–1; our translation.

79 Alfred Loisy, *Leçon d'ouverture du cours d'histoire des religions au Collège de France, 24 avril 1909*, Paris: Vrin, 1909, p. 5; our translation.

80 *Ibid.*, p. 27.

81 *Ibid.*, p. 28.

82 De Bujanda, *Index librorum prohibitorum*, pp. 556–7.

83 On the history of critical exegesis in the twentieth century, see Laplanche, *La Crise de l'origine*.

84 The few books that do mention this condemnation are quick to add that

this was only a regional Council that did not represent the entire Church even though it was not disavowed by Rome. See, for example, Dominique Lambert, 'Un acteur majeur de la réception du darwinisme à Louvain: Henry de Dorlodot', *Revue théologique de Louvain*, vol. 40, 2009, p. 505.

85 Cited in Jacques Arnould, *L'Église et l'histoire de la nature*, Paris: Cerf, 2000, p. 57; our translation.

86 Mariano Artigas, Thomas F. Glick and Rafael A. Martinez, *Negotiating Darwin: The Vatican Confronts Evolution, 1877–1902*, Baltimore, MD: Johns Hopkins University Press, 2007, p. 25.

87 Cited in *ibid.*, *Negotiating Darwin*, p. 36.

88 *Ibid.*, p. 47.

89 Leonard Alberstadt, 'Alexander Winchell's Preadamites: A Case for Dismissal from Vanderbilt University', *Earth Sciences History*, vol. 13, no. 2, 1994, pp. 97–112.

90 Ahmad Dallal, *Islam, Science and the Challenge of History*, New Haven, CT: Yale University Press, 2010, p. 165; the author writes 'Edmund Lewis' but it is in fact Edwin Lewis. For more details, see Anne-Laure Dupont, *Ǧurǧī Zaydān (1861–1914). Écrivain réformiste et témoin de la Renaissance arabe*, Damascus: Presses de l'IFPO, 2006, chapter 4, online: http://books. openedition.org/ifpo/5454.

91 David L. Livingstone, *Dealing with Darwin: Place, Politics and Rhetoric in Religious Engagement with Evolution*, Baltimore, MD: Johns Hopkins University Press, 2014, pp. 117–56.

92 Émile Ferrière, *Les Mythes de la Bible*, Paris: Félix Alcan, 1893, pp. 6–7; our translation.

93 Cited in Arnould, *L'Église et l'histoire de la nature*, p. 63; our translation.

94 Joseph Brucker, 'L'Origine de l'homme d'après la Bible et le transformisme', *Études*, vol. 47, 1889, p. 39; our translation.

95 *Ibid.*, p. 50.

96 *Ibid.*, p. 44.

97 Cited in Francesco Beretta, 'Les congrès scientifiques internationaux des catholiques (1888–1900) et la production d'orthodoxie dans l'espace intellectuel catholique', in Claude Langlois and Christian Sorrel (eds), *Le Catholicisme en Congrès, 2005. Chrétiens et sociétés. Documents et mémoires*, 2009, p. 13, online: https://halshs.archives-ouvertes.fr/halshs-00453294; our translation.

98 *Ibid.*, p. 15.

99 Cited by Arnould, *L'Église et l'histoire de la nature*, pp. 64–5.

100 Cited in Régis Ladous, 'Le magistère au défi de la modernité. Ou l'impossible distinction des sciences (1870–1920)', *Revue d'histoire ecclésiastique*, vol. 95, 2000, p. 652; our translation.

101 Cited in Scott Appleby, 'Between Americanism and Modernism: John Zahm and Theistic Evolution', *Church History*, vol. 56, 1987, p. 488.

102 Jean Rivière, review of Dorlodot's book, *Revue des sciences religieuses*, vol. 3, 1923, pp. 275–6; our translation.

103 Fanny Defrance-Jublot, 'Le darwinisme au regard de l'orthodoxie catholique. Un manuscrit exhumé', *Revue d'histoire des sciences humaines*, no. 22, 2010/1, pp. 229–37.

104 Arnaud Hurel, *L'Abbé Breuil. Un préhistorien dans le siècle*, Paris: CNRS éditions, 2014, pp. 277–85.

105 Cited in Raf De Bont, 'Rome and Theistic Evolutionism: The Hidden Strategies Behind the "Dorlodot Affair", 1920–1926', *Annals of Science*, vol. 62, 2005, p. 474.
106 The original manuscript has recently been found and published: Henry de Dorlodot, *L'Origine de l'homme. Le darwinisme au point de vue de l'orthodoxie catholique*, unpublished text presented and annotated by Marie-Claire Groessens-Van Dyck and Dominique Lambert, Collines de Wavre: Éditions Mardaga, 2009.
107 Cited in Defrance-Jublot, 'Le darwinisme au regard de l'orthodoxie catholique', p. 234; our translation.
108 Étienne Fouilloux, 'Un regain d'antimodernisme?', in Pierre Colin (ed.), *Intellectuels chrétiens et esprit des années 20*, Paris: Cerf, 1997, p. 97; see also Laplanche, *La Crise de l'origine*, pp. 138–9.
109 Cited in Fouilloux, 'Un regain d'antimodernisme?', p. 113; our translation.
110 Francis J. Wenninger, 'Catholicism and Catholic Thought by Canon Dorlodot', *The American Midland Naturalist*, vol. 8, 1923, pp. 211–14.
111 Cited in Dominique Lambert, 'Un acteur majeur de la réception du darwinisme à Louvain: Henry de Dorlodot', *Revue théologique de Louvain*, vol. 40, 2009, p. 518; our translation.
112 De Bont, 'Rome and Theistic Evolutionism', pp. 476–7.
113 François Azouvi, *La Gloire de Bergson. Essai sur le magistère philosophique*, Paris: Gallimard, 2007, p. 155.
114 Pierre Teilhard de Chardin, *Lettres à Édouard Le Roy (1921–1946)*, Paris: Éditions Facultés jésuites de Paris, 2008, letter of 21 December 1931, pp. 126–7; our translation.
115 Hurel, *L'Abbé Breuil*, pp. 318–20.
116 Cited in Régis Ladous, *Des Nobel au Vatican. La fondation de l'académie pontificale des sciences*, Paris: Cerf, 1994, p. 60; our translation.
117 *Ibid.*, p. 113.
118 Pius XII, *Humani Generis*, 12 August 1950, English text on http://w2.vatican.va/content/pius-xii/en/encyclicals/documents/hf_p-xii_enc_120819 50_humani-generis.html; our emphasis.
119 Paul-Émile Léger, *Les Origines de l'homme*, Montreal: Fides, 1961, p. 26; our translation.
120 Pope John Paul II, 'Message to the Pontifical Academy of Science on Evolution', 22 October 1996, https://www.ewtn.com/library/PAPALDOC/JP961022.HTM#note; emphasis added.
121 John Paul II, 'Address to the Plenary Session on "The Emergence of Complexity in Mathematics, Physics, Chemistry and Biology"', 31 October 1992, in *Papal Addresses to the Pontifical Academy of Sciences 1917–2002 and to the Pontifical Academy of Social Sciences, 1994–2002*, Vatican City: The Pontifical Academy of Sciences, Scripta Varia 100, 2003, p. 340.
122 Pope John Paul II, 'Message to the Pontifical Academy of Science on Evolution', 22 October 1996, emphasis added.
123 Galileo's Letter to the Grand Duchess Christina, in Maurice A. Finocchiaro (ed.) *The Galileo Affair. A Documentary History*, Berkeley, CA: University of California Press, 1989, p. 96.

5. From Conflict to Dialogue?

1 Friedrich Nietzsche, *Human, All Too Human, Parts One and Two*, translated by Helen Zimmern and Paul V. Cohn, Mineola, NY: Dover Publications, 2006, p. 71.

2 Ronald L. Numbers (ed.), *Galileo Goes to Jail and Other Myths about Science and Religion*, Cambridge, MA: Harvard University Press, 2009, p. 1; emphasis added. Note also the curious fact that after negating the obvious existence of many conflicts, the author then describes many of them in the period 1820 to 1860, many years before the creation of the 'myth' by Draper and White. Note also that this book 'would not exist' without the 'moral and financial backing' of the John Templeton Foundation (p. xi). As we will see later in this chapter, this Foundation played a central role in the growth of publications devoted to the demonstration of a 'dialogue' between science and religion and in minimizing the episodes of open and obvious conflicts by transforming them into 'meetings' and 'encounters'. It goes without saying that the Foundation gave the authors license 'to follow the evidence wherever it led' (*ibid.*).

3 *Ibid.*, p. 3; emphasis added.

4 *Ibid.*

5 Glenn C. Altschuler, 'From Religion to Ethics: Andrew D. White and the Dilemma of a Christian Rationalist', *Church History*, vol. 47, no. 3, 1978, pp. 308–24.

6 There are some exceptions: Ronald L. Numbers ironically notes that 'the Spanish edition *fittingly won a spot* on the Index of Prohibited Books', in his paper 'Aggressors, Victims and Peacemakers: Historical Actors in the Drama of Science and Religion', in Harold W. Attridge (ed.), *The Religion and Science Debate: Why Does It Continue?*, New Haven, CT: Yale University Press, 2009, p. 33; emphasis added. Note also that this collection of papers is the fruit of the *Terry Lectures*, the objective of which is to promote 'the *convergence of religion and science* for human welfare' (*ibid.*, p. 14, emphasis added).

7 George Lincoln Burr, 'Sketch of Andrew Dickson White', *Popular Science Monthly*, vol. 48, 1896, pp. 546–56; White explains the context in the introduction of his book, Andrew Dickson White, *A History of the Warfare of Science and Theology in Christendom*, New York: D. Appleton & Co., 1901 (1896).

8 Andrew Dickson White, 'The Warfare of Science', *Popular Science Monthly*, vol. 8, February 1876, pp. 385–409, and March 1876, pp. 553–70; for more detail, see David N. Livingstone, 'Re-Placing Darwinism and Christianity', in David C. Lindberg and Ronald L. Numbers (eds), *When Science and Christianity Meet*, Chicago, IL: University of Chicago Press, pp. 192–3.

9 Altschuler, 'From Religion to Ethics', pp. 315–16.

10 See, for example, the critical reviews of Draper's book in *Popular Science Monthly*, vol. 7, June 1875, pp. 230–3.

11 John William Dawson, 'The So-Called Conflict of Science and Religion', *Popular Science Monthly*, vol. 10, November 1876, p. 73, emphasis added. On Dawson, see Susan Sheets-Pyenson, *John William Dawson: Faith, Hope, and Science*, Montreal: McGill-Queen's University Press, 1995.

12 See, for example, Margaret J. Osler, 'Mixing Metaphors: Science and Religion or Natural Philosophy and Theology in Early Modern Europe', *History of Science*, vol. 35, 1997, pp. 91–113.

13 John Hedley Brooke, *Science and Religion: Some Historical Perspectives*, Cambridge: Cambridge University Press, 1991, pp. 34–5; for an analysis of the reception of White's book, see Altschuler, 'From Religion to Ethics', pp. 316–24.

14 John William Draper, *History of the Conflict between Religion and Science*, New York: D. Appleton & Co., 1875, p. x.

15 David C. Lindberg and Ronald L. Numbers, 'Beyond War and Peace: A Reappraisal of the Encounter Between Christianity and Science', *Church History*, vol. 55, 1986, pp. 338–54; the title itself clearly suggests that the authors have taken a position on the question of the relations between science and religion instead of analysing it historically. On the use of military metaphors in this debate, see James R. Moore, *The Post-Darwinian Controversies*, Cambridge: Cambridge University Press, 1979, pp. 19–49.

16 Marc Bloch, *L'Histoire, la Guerre, la Résistance*, Paris: Quarto, Gallimard, 2007, p. 947.

17 See Osler, 'Mixing Metaphors'.

18 Yuri Lin, Jean-Baptiste Michel, Erez Lieberman Aiden, Jon Orwant, Will Brockman and Slav Petrov, 'Syntactic Annotations for the Google Books Ngram Corpus', *Proceedings of the 50th Annual Meeting of the Association for Computational Linguistics*, Jeju, South Korea, 8–14 July 2012, pp. 169–74. Note that the vertical axis measures the relative proportion of the chosen words in the entire corpus of Google Books; the trend of the curve is thus not affected by the rise in the absolute number of books over the years.

19 For a more detailed analysis of this period, see Peter Bowler, *Reconciling Science and Religion: The Debate in Early-Twentieth-Century Britain*, Chicago, IL: University of Chicago Press, 2001.

20 Guillaume Carnino, *L'Invention de la science. La nouvelle religion de l'âge industriel*, Paris: Seuil, 2015, pp. 61–86.

21 Hélène Gispert (ed.), *Par la science, pour la patrie. L'Association française pour l'avancement des sciences (1872–1914). Un projet politique pour une société savante*, Rennes: Presses universitaires de Rennes, 2002.

22 The data are based on Harvard University Widener Library catalogue, well known to cover most published books. Worldcat could not be used as it includes many copies of the same book, but the general trend was the same.

23 The French corpus being more limited, frequency fluctuations are much larger and the analysis less reliable.

24 This book is mentioned by Ronald Numbers but its content is not analysed; see Numbers, 'Aggressors, Victims and Peacemakers', p. 16 as well as *Galileo Goes to Jail*, p. 3.

25 Thomas Dick, *The Christian Philosopher, or the Connection of Science and Philosophy with Religion*, Hartford, CT: D.F. Robinson & Co., 1833, p. 21; emphasis added.

26 *Ibid.*, p. 118; emphasis added.

27 Frederick Nolan, *The Analogy of Revelation and Science*, Oxford: J.H. Parker, 1833, p. xii, emphasis added.

28 *Ibid.*, p. 1.

29 *Ibid.*, pp. 3–4, emphasis added.

30 *Ibid.*, pp. 5–6.
31 *Ibid.*, p. viii.
32 Plato, *Laws*, Book XII, translated with an introduction by Benjamin Jowett, University of Adelaide EBooks, 2014, https://ebooks.adelaide.edu.au/p/plato/p71l/book12.html.
33 *The British Magazine*, 1 January 1834, 'Notices and Reviews', p. 72.
34 Michael Ruse, 'The Relationships Between Science and Religion in Britain, 1830–1870', *Church History*, vol. 44, 1975, p. 507.
35 T.S. Mackintosh, *The 'Electrical Universe' or The Elements of a Physical and Moral Philosophy*, Boston, MA: Josiah P. Mendum, 1846, p. 13; emphasis added. This is the American edition of the book first published in London in 1838.
36 'Critical Notices', *The North American Review*, vol. 45, No. 97, 1837, pp. 488–9, emphasis added.
37 John Henry Newman, 'Christianity and Physical Science. A Lecture in the School of Medicine', in *The Idea of University*, London: Longmans, Green & Co., 1907, p. 429; emphasis added.
38 *Ibid.*, p. 431.
39 *Ibid.*, p. 433.
40 *Ibid.*, p. 440.
41 *Ibid.*, pp. 440–1.
42 John Paul II, *Fides et Ratio*, 14 September 1998, http://w2.vatican.va/content/john-paul-ii/en/encyclicals/documents/hf_jp-ii_enc_14091998_fides-et-ratio.html.
43 White to George Lincoln Burr, 26 August 1885, cited in Altschuler, 'From Religion to Ethics', p. 315.
44 Ernest Renan, *The Future of Science, Ideas of 1848*, translated by Albert D. Vandam and C.B. Pitman, London: Chapman and Hall, 1891, p. 55.
45 Stephen Jay Gould, 'Nonoverlapping Magisteria', *Natural History*, vol. 106, March 1997, pp. 16–22.
46 Pierre Duhem, 'Physics of a Believer', in *The Aim and Structure of Physical Theory*, Princeton, NJ: Princeton University Press, 1991 [1954], p. 283. We have modified the translation to make it consistent with the original French that reads: 'deux jugements qui n'ont pas les mêmes termes et *qui ne portent pas* sur les mêmes objets', but which has been mistranslated as 'two judgements not having the same terms *but bearing on the same objects*', which makes no sense.
47 Friedrich Nietzsche, *Human, All Too Human, Parts One and Two*, translated by Helen Zimmern and Paul V. Cohn, Mineola, NY: Dover Publications, 2006, p. 71.
48 Renan, *The Future of Science*, p. 56.
49 American Scientific Affiliation, *Modern Science and Christian Faith, a symposium on the relationship of the Bible and Modern science by members of the American Scientific Affiliation*, Wheaton, IL: van Kampen Press, 1948.
50 See http://network.asa3.org/.
51 See the website of the organization: www.cis.org.uk/about-cis/.
52 See the website: www.ctns.org/.
53 See the list on their website: www.ctns.org/research.html.
54 For more details on Iqbal, see Nidhal Guessoum, *Islam et Science. Comment concilier le Coran et la science moderne*, Paris: Dervy, 2013, pp. 96–7.

55 From the 'Statement of Perspective' on the website of the journal: www. zygonjournal.org.
56 François Euvé, 'Science et mystique après la modernité', *Études*, vol. 394, no. 1, 2001, pp. 59–68; Françoise Champion, 'La croyance en l'Alliance de la science et de la religion dans les nouveaux courants mystiques et ésotériques', *Archives des sciences sociales des religions*, no. 82, April–June 1993, pp. 205–22.
57 See www.templetonprize.org/previouswinner.html#burhoe.
58 Gilles Kepel, *The Revenge of God. The Resurgence of Islam, Christianity, and Judaism in the Modern World*, University Park, PA: Penn University Press, 1993.
59 Ahmad Dallal, *Islam, Science and the Challenge of History*, New Haven, CT: Yale University Press, 2010, p. 169.
60 J. Brooks Flippen, *Jimmy Carter, the Politics of Family, and the Rise of the Religious Right*, Athens, GA: University of Georgia Press, 2011.
61 Jeffrey D. Schultz, John G. West Jr., and Iain MacLean (eds), *Encyclopedia of Religion in American Politics*, Phoenix, AZ: The Oryx Press, 1999, p. 31.
62 Nicolas A. Rupke, 'Five Discourses of Bible and Science, 1750–2000', in Jed Z. Buchwald (ed.), *A Master of Science History: Essays in Honor of Charles Coulston Gillespie*, Dordrecht: Springer, 2012, pp. 179–80.
63 www.aaas.org/DoSER.
64 Sunny Bains, 'Questioning the Integrity of the John Templeton Foundation', *Evolutionary Psychology*, vol. 9, 2011, p. 99.
65 Paul Mitchell, 'Behind the Creationism Controversy at Britain's Royal Society', *World Socialist Website*, 17 October 2008, www.wsws.org/en/articles/2008/10/crea-o17.html.
66 Jack Morrell and Arnold Thackray, *Gentlemen of Science: Early Years of the British Association for the Advancement of Science*, Oxford: Oxford University Press, 1981, pp. 224–45; see also Stephen Gaukroger, *The Emergence of Scientific Culture. Science and the Shaping of Modernity*, Oxford: Oxford University Press, 2006, pp. 220–7.
67 Kathryn Pritchard, 'Religion and Science Can Have a True Dialogue', *Nature*, vol. 537, 22 September 2016, p. 451.
68 http://www.scientistsincongregations.org/about-us/.
69 *Nature* to Alan Sokal, 7 October 2016; thanks to Alan Sokal and Jerry Coyne for giving me permission to quote their email.
70 Emilio Chuvieco, Marcelo Sánchez Sorondo and Josef Settele, 'Religion and Science: Boost Sustainability', *Nature*, vol. 538, 27 October 2016, p. 459; David Lovelace, 'Religion and Science: Not a True Dialogue', *ibid*.
71 www.ctns.org/book_prize.html.
72 Dean Nelson and Karl Giberson, *Quantum Leap: How John Polkinghorne Found God in Science and in Religion*, Oxford: Monarch Books, 2011.
73 John Horgan, 'The Templeton Foundation: A Skeptic's Take,' *Edge*, 4 April 2006, online: http://edge.org/conversation/the-templeton-foundation-a-skeptic-39s-take.
74 www.templeton.org/what-we-fund/grants/courses-on-the-relation-among-science-religion-and-philosophy-from-an-orthodox-c.
75 www.templeton.org/what-we-fund/grants/jewish-tradition-and-the-challenge-of-darwinism.
76 See Guessoum, *Islam et science*.

77 Geoffrey Cantor, 'What Shall We Do with the "Conflict Thesis"', in Thomas Dixon, Geoffrey Cantor and Stephen Pumphrey (eds), *Science and Religion: New Historical Perspectives*, Cambridge: Cambridge University Press, 2010, pp. 283–98.

78 The Templeton Foundation, 'A Chronicle. Prof. Charles Taylor. 2007 Templeton Prize Laureate', http://www.templetonprize.org/pdfs/Templeton_Prize_Chronicle_2007.pdf, p. 13; the same sentence can be found in newspapers of the time: Guillaume Bourgault-Côté, 'Charles Taylor honoré par la Fondation Templeton', *Le Devoir*, 15 March 2007. It is notable that such sentiments are largely absent from Taylor's substantive work on secularism, *A Secular Age*, Cambridge, MA: Harvard University Press, 2007.

79 Article by Abbé Maret published in 1834 in *L'Univers*, cited in François Laplanche (ed.), *La science catholique. L'encyclopédie catholique de Migne (1844–1873) entre apologétique et vulgarisation*, Paris: Cerf, 1992, p. 41.

80 Frère Marie-Victorin, *Science, culture et nation*, Yves Gingras (ed.), Montreal: Boréal, 1996, p. 85.

81 Robert N. Proctor and Londa Schiebinger (eds) *Agnotology: The Making and Unmaking of Ignorance*, Stanford, CA: Stanford University Press, 2008.

82 Ronald L. Numbers, 'Simplifying Complexity: Patterns in the History of Science and Religion', in Dixon, Cantor and Pumphrey (eds), *Science and Religion*, p. 274.

83 Thomas Dixon, 'Introduction', in *ibid.*, p. 5.

84 *Ibid.*, p. xiv.

85 http://www.discovery.org/about.

86 Randy Moore, Mark Decker and Sehoya Cotner, *Chronology of the Evolution-Creation Controversy*, Santa Barbara, CA: Greenwood Press, 2010, p. 353.

87 *Ibid.*, pp. 360–1.

88 Dallal, *Islam, Science and the Challenge of History*, p. 169; Martin Riexinger, 'Turkey', in Stefaan Blancke, Hans Henrik Hjermitslev and Peter J. Kjaergaard (eds), *Creationism in Europe*, Baltimore, MD: Johns Hopkins University Press, 2014, pp. 180–98. For previous debates, see Aldel Ziadat, *Western Science in the Arab World: The Impact of Darwinism, 1860–1930*, New York: St. Martin's Press, 1986; Marwa Elshakry, *Reading Darwin in Arabic, 1860–1950*, Chicago, IL: University of Chicago Press, 2014; Farid El Asri, 'Discours musulman et sciences modernes: un état de la question', in Brigitte Maréchal and Félice Dassento (eds), *Adam et l'évolution. Islam et christianisme confrontés aux sciences*, Louvain-la-Neuve: Academia Bruylant, 2009, pp. 109–23; B. Harun Küçük, 'Islam, Christianity, and the Conflict Thesis', in Dixon, Cantor and Pumphrey (ed.), *Science and Religion*, pp. 111–30.

89 Fouzia Farida Charfi, *La Science voilée*, Paris: Odile Jacob, 2013, p. 16.

6. What Is a 'Dialogue' Between Science and Religion?

1 Ernest Renan, *The Future of Science, Ideas of 1848*, translated by Albert D. Vandam and C.B. Pitman, London: Chapman and Hall, 1891, p. 56.

2 http://www.oxforddictionaries.com/definition/english/argument.

3 For a detailed study of dialogues, see Marc Angenot, *Dialogues de sourds. Traité de rhétorique antilogique*, Paris: Mille et une nuits, 2008; and Yves

Gingras (ed.), *Controverses. Accords et désaccords en sciences humaines et sociales*, Paris: CNRS, 2014.

4 Alexandre Koyré, *Discovering Plato*, translated by Leonora Cohen Rosenfield, New York: Columbia University Press, 1960, p. 16.

5 John Henry Newman, 'Christianism and Physical Science, Lecture in the School of Medicine', *The Idea of University*, London: Longmans, Green & Co., 1907, p. 431.

6 I discuss miracles in science in greater detail in Yves Gingras, *Parlons Sciences. Entretiens avec Yanick Villedieu sur les transformations de l'esprit scientifique*, Montréal: Boréal, 2008, pp. 213–18.

7 John D. Barrow and Frank J. Tipler, *The Anthropic Cosmological Principle*, Oxford: Oxford University Press, 1986; Franck J. Tipler, *The Physics of Immortality: Modern Cosmology, God and the Resurrection of the Dead*, New York: Anchor Books, 1994.

8 Véronique Le Ru, *La Nature, miroir de Dieu*, Paris: Vuibert, 2010, pp. 81–2.

9 Martin Rees, *Before the Beginning: Our Universe and Others*, Reading, MA: Perseus Books, 1997, p. 243.

10 *Le monde s'est-il créé tout seul? Entretiens avec Patrice van Eersel, avec la collaboration de Sylvain Michelet*, Paris: Albin Michel, pp. 88–90.

11 John Horgan, 'The Templeton Foundation: A Skeptic's Take', *Edge*, 4 April 2006, online: http://edge.org/conversation/the-templeton-foundation-a-skeptic-39s-take; emphasis added.

12 National Academy of Sciences, Working Group on Teaching Evolution, *Teaching About Evolution and the Nature of Science*, Washington, DC: National Academy Press, 1998, pp. 43 and 58.

13 Bernadette Bensaude-Vincent and Christine Blondel (eds), *Des savants face à l'occulte, 1870–1940*, Paris: La Découverte, 2002; Sal P. Restivo, 'Parallels and Paradoxes in Modern Physics and Eastern Mysticism: I. A Critical Reconnaissance', *Social Studies of Science*, vol. 8, no. 2, 1978, pp. 143–81; Sal P. Restivo, 'Parallels and Paradoxes in Modern Physics and Eastern Mysticism: II. A Sociological Perspective on Parallelism', *Social Studies of Science*, vol. 12, no. 1, 1982, pp. 37–71.

14 Gaston Bachelard, *The Formation of the Scientific Mind. A Contribution to a Psychoanalysis of Objective Knowledge*, translated by Mary McAllester Jones, Manchester: Clinamen Press, 2002, p. 26.

15 Wiktor Stoczkowski, *Des hommes, des dieux et des extraterrestres. Ethnologie d'une croyance moderne*, Paris: Flammarion, 1999.

16 David Kaiser, *How the Hippies Saved Physics: Science, Counterculture and the Quantum Revival*, New York: W.W. Norton, 2012, p. 168.

17 On the reception of Capra's book, see Kaiser, *ibid.*, pp. 153–65.

18 Fritjof Capra, *The Tao of Physics*, Second Edition, Boston, MA: New Science Library, 1985, p. 25.

19 Ilya Prigogine and Isabelle Stengers, *La Nouvelle Alliance*, Paris: Gallimard, 1979, translated as *Order Out Of Chaos: Man's New Dialogue with Nature*, New York: Bantam New Age Books, 1984.

20 Henri Atlan, *Croyances. Comment expliquer le monde?*, Paris: Autrement, 2014, pp. 21–2; for a critical analysis of these currents, see Dominique Terré-Fornacciari, *Les Sirènes de l'irrationnel. Quand la science touche à la mystique*, Paris: Albin Michel, 1991; John Horgan, *Rational Mysticism:*

Dispatches from the Border Between Science and Spirituality, Boston, MA: Houghton Mifflin, 2003.

21 Michael Riordan, Lillian Hoddeson and Adrienne W. Kolb, *Tunnel Visions. The Rise and Fall of the Superconducting Super Collider*, Chicago, IL: University of Chicago Press, 2015.

22 Stephen Hawking, *A Brief History of Time*, New York: Bantam, 1988, p. 173.

23 Francis S. Collins, *The Language of God: A Scientist Presents Evidence for Belief*, New York: Free Press, 2007.

24 *Le Nouvel Observateur*, 21–27 December 1989, p. 13.

25 For a detailed analysis of the esoteric movements of the 1960s, see Stoczkowski, *Des hommes, des dieux et des extraterrestres*.

26 www.templetonprize.org/purpose.html; on the change of name, see Sunny Bains, 'Questioning the Integrity of the John Templeton Foundation', *Evolutionary Psychology*, vol. 9, 2011, p. 94.

27 Charles H. Townes, 'The Convergence of Science and Religion', *Think*, vol. 32, March–April 1966, pp. 2–7.

28 www.templetonprize.org/previouswinner.html.

29 For a detailed critique of the myth of 'fine tuning', see Victor J. Steinger, *The Fallacy of Fine-Tuning: Why the Universe Is Not Designed for Us*, New York: Prometheus Books, 2011; Mark Colyvan, Jay L. Garfield and Graham Priest, 'Problems with the Argument from Fine Tuning', *Synthese*, vol. 145, no. 3, 2005, pp. 325–38.

30 George Ellis, 'The Anthropic Principle: Laws and Environments', in F. Bertola and U. Curi (eds), *The Anthropic Principle*, Cambridge: Cambridge University Press, 1993, p. 30.

31 George Ellis, 'Piety in the Sky', *Nature*, vol. 371, 8 September 1994, p. 115.

32 www.templetonprize.org/previouswinner.html; emphasis added.

33 Jerry A. Coyne, 'Martin Rees and the Templeton Travesty', *The Guardian*, 6 April 2011; see also Jerry A. Coyne, *Faith vs. Fact: Why Science and Religion are Incompatible*, New York: Viking, 2015.

34 www.templetonprize.org/previouswinner.html.

35 Sunny Bains, 'Questioning the Integrity of the John Templeton Foundation'.

36 For more information on the UIP, see Cyrille Baudouin and Olivier Brosseau, *Les Créationnismes. Une menace pour la société française?*, Paris: Syllepse, 2008, pp. 45–53; and, by the same authors, *Enquête sur les créationnismes. Réseaux, stratégies et objectifs politiques*, Paris: Belin, 2013.

37 www.templetonpress.org/content/science-and-search-meaning.

38 Jean Staune (ed.), *Science and the Search for Meaning*, West Conshohocken, PA: Templeton Foundation Press, 2006, p. 6.

39 www.templeton.org/what-we-fund/grants/science-and-religion-in-islam.

40 www.templeton.org/what-we-fund/grants/science-and-orthodoxy; see also the website www.it4s.ro/.

41 www.templeton.org/what-we-fund/grants/science-and-islam-an-educational-approach.

42 Nidhal Guessoum, *Islam's Quantum Question. Reconciling Muslim Tradition and Modern Science*, London: I.B. Tauris, 2011. The content also appeared in French in two volumes: *Islam et Science. Comment concilier le Coran et la science moderne* (Paris: Dervy, 2013) and *Islam, Big-bang et Darwin. Les questions qui fâchent* (Paris: Dervy, 2015). Note that Guessoum is opposed

to simplistic approaches like those promoted by Maurice Bucaille, who published in the mid-1970s the best-selling *The Bible, The Qu'ran and Science*, in which he claims that all the great scientific discoveries can already be found in those religious texts.

43 www.templeton.org/what-we-fund/grants/the-issr-science-and-religion-library-initiative.

44 www.templeton.org/what-we-fund/grants/templeton-cambridge-journalism-fellowships-and-seminars-in-science-and-religion.

45 See, for example, https://www.templeton.org/templeton_report/20080611/.

46 www.templeton.org/what-we-fund/grants/templeton-cambridge-journalism-fellowships-and-seminars-in-science-and-religion.

47 The texts of these conferences have been published in Russell Re Manning and Michael Byrne (eds), *Science and Religion in the Twenty-First Century*, London: SCM Press, 2013.

48 Benedict XVI, 'Faith, Reason and the University. Memories and Reflections', *Aula Magna* of the University of Regensburg, Tuesday, 12 September 2006, https://w2.vatican.va/content/benedict-xvi/en/speeches/2006/september/docu ments/hf_ben-xvi_spe_20060912_university-regensburg.html. On the opposition between Scotus and Aquinas, see Yves Gingras, 'Duns Scot vs Thomas d'Aquin: le moment québécois d'un conflit multi-séculaire', *Revue d'histoire de l'Amérique française*, vol. 62, no. 3–4, 2009, pp. 377–406.

49 Michel Benoît, *Naissance du Coran. Aux origines de la violence*, Paris: L'Harmattan, 2015.

50 Voltaire, *Toleration and other essays*, Joseph McCabe (ed.), New York: G.P. Putnam's Sons, 1912, p. 259.

51 Chris Hastings, Patrick Hennessy and Sean Rayment, 'Archbishop of Canterbury: This Has Made Me Question God's Existence', *The Telegraph*, 2 January 2005.

52 *Le Monde*, 25 September 2015; our translation; see also https://en.wikipedia. org/wiki/2015_Mina_stampede.

53 For a critical analysis of this conference, see Jean Bollack, Christian Jambert and Abdelwahab Medded, *La Conférence de Ratisbonne. Enjeux et controverses*, Paris: Bayard, 2007.

54 Benedict XVI, 'Faith, Reason and the University'.

55 Joseph Ratzinger, 'Démocratie, droit et religion', *Esprit*, July 2004, p. 23.

56 http://w2.vatican.va/content/francesco/en/encyclicals/documents/papa-franc esco_20150524_enciclica-laudato-si.html, paragraphs, 62, 216 and 217.

7. Belief Versus Science

1 Arthur Schopenhauer, 'On religion', in *Essays and Aphorisms*, edited and translated by R.J. Hollingdale, London: Penguin Classics, 1981 (1970), p. 180.

2 Gilles Kepel, *The Revenge of God. The Resurgence of Islam, Christianity, and Judaism in the Modern World*, University Park, PA: Penn State University Press, 1993.

3 Peter R. Afrasiabi, 'Property Rights in Ancient Skeletal Remains', *Southern California Law Review*, vol. 70, 1997, pp. 805–39.

4 Natalie Angler and John Dunn, 'Burying Bones of Contention', *Time*, vol. 124, 10 September 1984, p. 36.

5 Virginia Morrell, 'Who Owns the Past?', *Science*, vol. 268, June 9, 1995, pp. 1425–6.
6 *Ibid.*, p. 1424.
7 *Ibid.*, p. 1424.
8 *Ibid.*, p. 1425.
9 J.S. Cybulski, N.S. Ossenberg and W.D. Wade, 'Committee Report: Statement on the Excavation, Treatment, Analysis and Disposition of Human Skeletal Remains from Archaeological Sites in Canada', *Canadian Review of Physical Anthropology*, vol. 1, no. 1, 1979, p. 36.
10 Max Weber, *The Methodology of the Social Sciences*, Glencoe, IL: Free Press, 1949, p. 110.
11 Douglas H. Uberlaker and Lauryn Guttenplan Grant, 'Human Skeletal Remains: Preservation or Reburial?', *Yearbook of Physical Anthropology*, vol. 32, 1989, pp. 249–87.
12 David H. Thomas, *Skull Wars*, New York: Basic Books, 2000, pp. 214–15.
13 Virginia Morell, 'An Anthropological Culture Shift', *Science*, vol. 264, 15 March 1994, p. 20.
14 *Ibid.*, p. 20.
15 *Ibid.*, p. 21.
16 See, for example, Constance Holden, 'UCSD Back Off On Bone Bid', *Science*, vol. 324, 17 April 2009, p. 317; Jef Akst, 'Bones Won't Be Buried Yet', *The Scientist*, 10 May 2012; Helen Shen, 'Ancient Bones Stay Put for Now in California Lawsuit', *Nature*, 8 May 2012.
17 Andrew Lawler, 'A Tale of Two Skeletons', *Science*, vol. 330, 8 October 2010, p. 171.
18 Thomas, *Skull Wars*, p. 239.
19 Martin Kaste, '"Kennewick" Case Fails to End Battle over Bones', NPR, 12 June 2005, online at: www.npr.org/templates/story/story.php?storyId=4699997.
20 Ernest Renan, *The Future of Science, Ideas of 1848*, translated by Albert D. Vandam and C.B. Pitman, London: Chapman and Hall, 1891, p. 37.
21 Marc Angenot, *Dialogues de sourds. Traité de rhétorique antilogique*, Paris: Mille et une nuits, 2008; Yves Gingras (ed.), *Controverses. Accords et désaccords en sciences sociales et humaines*, Paris: CNRS, 2014.
22 James C. Chatters, 'The Recovery and First Analysis of an Early Holocene Human Skeleton from Kennewick', *Society for American Archeology*, vol. 65, no. 2, 2000, pp. 291–316.
23 For more details, see Thomas, *Skull Wars*.
24 Virginia Morell, 'A Tangled Affair of Hair and Regulations', *Science*, vol. 268, 9 June 1995, p. 1425; 'Pulling Hair from the Ground', *Science*, vol. 265, 5 August 1994, p. 741.
25 *Bonnichsen v. United States*, US Court of Appeals, Ninth Circuit, Nos 02-35996, 2004, Section IV.
26 *Ibid.*
27 www.burkemuseum.org/kman/.
28 *Bonnichsen v. United States*, Introduction.
29 Brief Amicus Curia of Pacific Legal Foundation in Support of Plaintiffs-Appellees Robson Bonnichsen *et al.*, No. 02-35996, United States Court of Appeals for the Ninth Circuit, 2003.

30 Don Sampson, 'Ancient One/Kennewick Man: Tribal Chair Questions Scientists' Motives and Credibility', 21 November 1997, https://introduc tiontoarchaeology.files.wordpress.com/2013/11/ancient-one-motives-and-credibility-kman2.pdf.

31 Matthew Preusch, 'Judge Backs Study of Ancient Human Remains', *The New York Times*, 5 February 2004, p. A-20.

32 George Johnson, 'Indian Tribes' Creationists Thwart Archeologists', *The New York Times*, 22 October 1996, p. C-13.

33 Thomas, *Skull Wars*, pp. 242–3.

34 H.H. Gerth and C. Wright Mills (eds), *From Max Weber: Essays in Sociology*, New York: Oxford University Press, 1946, p. 152.

35 Douglas W. Owsley and Richard L. Jantz, 'Archaeological Politics and Public Interest in Paleoamerican Studies: Lessons from Gordon Creek Woman and Kennewick Man', *American Antiquity*, vol. 66, 2001, pp. 565–75.

36 Lynda V. Maypes, 'Kennewick Man Bones not from Columbia Valley, Scientist Tells Tribes', *The Seattle Times*, 9 October 2012, http://www.seattletimes.com/seattle-news/kennewick-man-bones-not-from-columbia-val ley-scientist-tells-tribes/.

37 Carl Zimmer, 'New DNA Results Show Kennewick Man Was Native American', *The New York Times*, 18 June 2015.

38 Morten Rasmussen et al., 'The Ancestry and Affiliations of Kennewick Man', *Nature*, vol. 523, 23 July 2015, pp. 455–8.

39 News in Brief, '"Ancient One" to Get Native American Burial', *Science*, vol. 352, 6 May 2016, p. 634.

40 *Bonnichsen v. United States*.

41 Constance Holden, 'U.S. Government Shifts Stance on Claims to Ancient Remains', *Science*, vol. 309, 5 August 2005, p. 861; Constance Holden, 'Remains Remain Controversial', *Science*, vol. 318, 19 October 2007, p. 377; Susan B. Bruning, 'Complex Legal Legacies: The Native American Graves Protection and Repatriation Act, Scientific Study, and Kennewick Man', *American Antiquity*, vol. 71, July 2006, pp. 501–21.

42 Douglas W. Owsley and Richard L. Jantz (eds), *Kennewick Man: The Scientific Investigation of an Ancient American Skeleton*, College Station, TX: Texas A&M University Press, 2014. See also Douglas Preston, 'The Kennewick Man Finally Freed to Share His Secrets', *Smithsonian Magazine*, September 2014.

43 Keith Kloor, 'Giving Back the Bones', *Science*, vol. 330, 8 October 2010, p. 166.

44 The question is not purely rhetorical as astronomers have recently discov-ered. They have seen their latest project of constructing a new telescope on Mauna Kea, in Hawaii – where thirteen of them have been put in place since the 1970s – stopped by a group Native Hawaiians who consider it to be sacred land; see Alexandra Witze, 'Mountain Battle', *Nature*, vol. 526, 1 October 2015, pp. 24–8; Toni Feder, 'The Future of Astronomy in Hawaii Hinges on the Thirty Meter Telescope', *Physics Today*, July 2016, pp. 31–3.

45 Paul A. Offit, *Bad Faith: When Religious Belief Undermines Modern Medicine*, New York: Basic Books, 2015, pp. 193–4; see also Marci A. Hamilton, *God vs. the Gavel: The Perils of Extreme Religious Liberty*, New York: Cambridge University Press, 2014.

46 For a synthesis of these results, see Janna C. Merrick, 'Spiritual Healing, Sick Kids and the Law: Inequities in the American Healthcare System', *American Journal of Law and Medicine*, vol. 29, 2003, p. 203; details on State exemptions are available at http://childrenshealthcare.org/?page_id=24#Exemptions. Thanks to Alan Sokal for pointing me towards that source.

47 Seth M. Asser and Rita Swan, 'Child Fatalities from Religion-Motivated Medical Neglect', *Pediatrics*, vol. 101, no. 4, 1998, pp. 625–9; see Table 2, p. 627.

48 *Ibid.*, Table 4, p. 628.

49 Offit, *Bad faith*, p. 184.

50 *Supreme Court of Canada, A.C. v. Manitoba (Director of Child and Family Services)*, 2009, SCC 30, p. 190.

51 Offit, *Bad Faith*, p. 101.

52 *Ibid.*, p. 111.

53 *Ibid.*, pp. 105–6.

54 *Ibid.*, p. 192.

55 *Ibid.*, p. 101.

56 Alyshah Hasham, 'Aboriginal Medicine Ruling Sparks Instant Controversy', *Toronto Star*, 14 November 2014.

57 Ontario Court of Justice, Brantford, Ontario, Judge G.B. Edward J., *Hamilton Health Sciences Corp. v. D.H.*, 14 November 2014, paragraph 81, our emphasis.

58 Kelly Grant, 'Aboriginal Girl Now Receiving both Chemo and Traditional Medicine', *The Globe and Mail*, 24 April 2015, http://www.theglobeandmail.com/news/national/case-over-cancer-treatment-for-native-girl-is-resolved/article24101800/; Joanna Frketich, 'Aboriginal Girl's Chemo Case Returns to Court Friday', *Hamilton Spectator*, 23 April 2015.

Conclusion: Betting on Reason

1 Georges Canguilhem, 'La philosophie d'Hermann Keyserling', *Libres Propos*, no. 1, 20 March 1927, in *Œuvres complètes*, vol. 1, *Écrits philosophiques et politiques 1926–1939*, Paris: Vrin, 2011, p. 156; my translation. Thanks to Camille Limoges for pointing out that text to me.

2 Yves Gingras, *Éloge de l'homo techno-logicus*, Montreal: Fides, 2005.

3 Henk van den Belt, 'Playing God in Frankenstein's Footsteps: Synthetic Biology and the Meaning of Life', *Nanoethics*, vol. 3, no. 3, 2009, pp. 257–68.

4 Robert K. Merton, *The Sociology of Science*, Chicago, IL: University of Chicago Press, 1973.

5 Alan Beyerchen, *Scientists under Hitler: Politics and the Physics Community in the Third Reich*, New Haven, CT: Yale University Press, 1977; Dominique Lecourt, *Proletarian Science? The Case of Lysenko*, translated by Ben Brewster, London: NLB, 1977; François Laplanche, 'La notion de science catholique: ses origines au début du XIXe siècle', *Revue d'histoire de l'Église de France*, vol. 74, no. 192, 1988, pp. 63–90; Taner Edis, *An Illusion of Harmony: Science and Religion in Islam*, Amherst, NY: Prometheus Books, 2007.

6 Ernest Renan, *The Future of Science, Ideas of 1848*, translated by Albert D. Vandam and C.B. Pitman, London: Chapman and Hall, 1891, p. 41.

7 François-David Sebbah, *Qu'est-ce que la technoscience? Une thèse épisté-mologique ou la fille du diable?*, Paris: Encre Marine, 2010; see also Dominique Raynaud, 'Note historique sur le mot "technosciènce"', 4 April 2015, online: http://zilsel.hypotheses.org/1875.

8 Thomas Kuhn, *The Structure of Scientific Revolutions*, second edition, Chicago, IL: University of Chicago Press, 1970; Imre Lakatos, *Proofs and Refutations. The Logic of Mathematical Discovery*, Cambridge: Cambridge University Press, 1976.

9 Jacques Bouveresse, *Peut-on ne pas croire? Sur la vérité, la croyance et la foi*, Paris: Agone, 2012.

10 Joseph Ratzinger, 'Démocratie, droit et religion', *Esprit*, July 2004, p. 28; our translation.

INDEX